Praise for [illegible]

“Any new book by William Zinsser [illegible] is latest, *Writing Places*, delights with lovely insights and compelling turns of phrase.” —*Parade*

“*Writing Places* is not a long book, but it possesses all the qualities that Zinsser believes matter most in good writing—clarity, brevity, simplicity and humanity.” —*Pittsburgh Post-Gazette*

“William Zinsser is a born teacher—and even more important to readers of books, a born writer. There are so many insights about the writing life and about the interesting places his life as a writer took him in this book, you come away refreshed, awed—and grateful.” —Thomas Fleming, author of *The Intimate Lives of the Founding Fathers*

“Zinsser’s characteristic good humor and conversational tone are present as he describes his numerous changes of job descriptions and employers, all while he pursued the same underlying vocation of communicating—and teaching others to communicate—via the written word.” —*Library Journal*

“William Zinsser turns his zest, warmth, and curiosity—his sharp but forgiving eye—on his own story. The result is lively, funny and moving, especially for anyone who cares about the art and business of writing well.” —Evan Thomas, *Newsweek*

“Superb . . . The book tops out at about 200 pages; I wanted more.” —Steve Weinberg, *The Writer*

ABOUT THE AUTHOR

As a writer, editor and teacher, WILLIAM ZINSSER continues to be a mentor to countless people who want to write with clarity and confidence. His 18 books include *On Writing Well,* which has sold almost 1.5 million copies. He lives in New York with his wife, Caroline Zinsser, and teaches at the New School and the Columbia Graduate School of Journalism.

Also by William Zinsser

BOOKS IN PRINT

On Writing Well

Writing to Learn

Mitchell & Ruff

American Places

Spring Training

Easy to Remember:
The Great American Songwriters and Their Songs

Writing About Your Life

Writing Places

EDITED BY WILLIAM ZINSSER

Inventing the Truth: The Art and Craft of Memoir

Worlds of Childhood:
The Art and Craft of Writing for Children

AUDIO

On Writing Well: CD Audio Collection
("On Writing Well" and "How to Write a Memoir")

WRITING PLACES

THE LIFE JOURNEY OF A WRITER AND TEACHER

William Zinsser

HARPER

NEW YORK • LONDON • TORONTO • SYDNEY

HARPER

A hardcover edition of this book was published in 2009 by Harper, an imprint of HarperCollins Publishers.

Portions of this book first appeared in the *New York Times* and the *Yale Alumni Magazine*. Chapter 2 first appeared in *The American Scholar* ("The Daily Miracle"), as did excerpts from chapters 13 and 14 ("Visions and Revisions").

William Zinsser's papers are at the Fales Library at New York University, located in the Elmer Bobst Library, 70 Washington Square South, New York, N.Y. 10012–1091. All the books and articles cited or excerpted in these pages are collected there, including the complete movie reviews from the *New York Herald Tribune* from 1954 to 1958.

FIRST HARPER PAPERBACK PUBLISHED 2010.

Designed by Eric Butler

The Library of Congress has catalogued the hardcover edition as follows:
Zinsser, William Knowlton.
Writing places : the life journey of a writer and teacher / William Zinsser.
p. cm.
ISBN: 978-0-06-172902-7
1. Zinsser, William Knowlton—Authorship. 2. Journalists—United States—Biography. 3. English teachers—United States—Biography. 4. Authors, American—20th century—Biography. I. Title.

PS3576.I5625Z479 2009
8089'.0092—dc22
[B] 2008044781

ISBN 978-0-06-172903-4 (pbk.)

HB 04.04.2024

Contents

1. Office for Rent 1

2. Herald Tribune Days 5

3. The Loneliness of the Freelance Writer 23

4. Travels with My Olivetti 29

5. Up in the Billiard House 39

6. Writing About the Sixties 49

7. Making a New Life 63

8. "Leaders in Their Generation" 69

9. Old Yale and New Yale 75

10. Becoming a Teacher 83

11. A Summons from the King 93

12. College Master 101

13. "On Writing Well" 119

14. Taking It on the Road 135

15. Processing Words 155

16. Only in New York 169

Acknowledgments 193

Writing Places

1

Office for Rent

Of all the places where I've done my writing, none was more unusual than the office that had a fire pole. The year was 1987, and I had just left an editor's job in New York to get back to my own writing and teaching. I'm not a writer who craves silence and solitude. I wanted to rent space from a larger company that had congenial people and a copying machine that worked—perhaps a firm of architects. Nosing about in the classified ads, I saw the words "East 50s publisher seeks subtenant." I called to inquire and learned that the publisher was Bernard Geis.

Bernard Geis! The name reverberated in my memory grooves. Nobody in the modern publishing industry had so subverted its historic image as a gentlemen's club, a world where authors looked like people who actually spent the day writing books. Geis's breakthrough books were Helen Gurley Brown's *Sex and the Single Girl* and Jacqueline Susann's *Valley of the Dolls*, which would spend 65 weeks on the *New York Times* best-seller list in the mid-1960s. His strenuous exertions turned both women into pop icons and certified the bedroom as literature's favorite playpen.

I found Bernard Geis Associates on the top two floors of a five-story office building at 128 East 56th Street. The elevator only went as far as the fourth floor, where a receptionist told me I would have to walk up one flight to meet Mr. Geis. She was the firm's sole remaining employee on that floor, which in its heyday was a hive of aggressive editors, marketers and publicists. Geis was the first publisher to saturate television with his authors; at one point it was almost impossible to turn on a TV set and not see Jacqueline Susann. Now the empire was dead—the firm filed for bankruptcy in 1971—and all those fourth-floor offices were rented out to freelancers in fields like fashion and design. Presumably that's where I would also end up.

Climbing to the fifth floor, I found myself in a light and airy room with angled ceilings that had the look of a Parisian garret. A balcony with window boxes full of geraniums ran around the outside. In the middle of the floor I noticed a large circular hole, through which a fire pole extended to the ceiling; below, it ended on the fourth floor, next to the receptionist. It looked like a long way down.

"Mr. Geis had that pole installed," I was told by Alice Baer, who, as office manager, was the firm's only other survivor, "and he always uses it when he leaves." Geis was then 78. Baer's office was also on that floor—she handled the firm's remaining rights and royalties—and so was the office she was offering me. It was a large and pleasant room, perfect for my needs, and we soon arrived at a rental agreement, subject only to my approval by the boss. He had the largest office, and I was ushered in to meet him.

Geis's walls were hung with the framed book jackets of his early blockbusters, including *Groucho and Me*, by Groucho Marx; *Kids Say the Darnedest Things*, by Art Linkletter; and *The Cardinal Sins*, by Father Andrew Greeley, whose novels contained more sex than Catholic priests were assumed to know about. Several other titles by Father Greeley were also on display, their jackets in various gradations of red, their titles embossed in raised gold letters. Obviously Greeley & Geis had been a profitable merger of church and state.

I half expected Geis to be a furtive figure, hunched over his desk. But he was entirely presentable—a short, well-dressed man with a full head of curly white hair and an eager-to-please smile. We sat and talked about books and authors, and I gave my best impersonation of a model tenant, a man whose checks wouldn't bounce and who wouldn't cause any fuss. Geis seemed ready to close the deal.

Then he said, "Have you been down the fire pole?" I told him I hadn't. I didn't say that the last thing I wanted to do was go down the fire pole. It looked shiny and slippery. I glanced at Geis to assure myself that he was kidding. He wasn't. The pole seemed to be some final test of renter suitability.

"What you do," he said, leading me to the abyss, "is grip the pole with your knees. That slows you down." I had never thought of myself as a man with strong knees. "You'll like it,"

Geis said. "Once you've done it you won't want to go down any other way."

I was taken back to all the times when I've stood over some body of cold water while being told how much I would like it if I would only just jump in. Often it was easy enough to demur. But I also remembered that life delivers a certain number of moments when we are finally required to jump.

I grabbed the pole, flung my body into the void, and felt myself in the throes of rapid descent. "Use your knees!" Geis shouted, but nothing slowed me down, and I landed with a thump that rattled my muscular-skeletal system and every organ it enclosed.

"Wasn't that great?" Geis called down.

"Great!" I called back. "I'll be right up to sign the lease."

2

Herald Tribune Days

I did my first writing as an adult in North Africa and Italy during World War II. That's where I learned that writers can write anywhere. Some writers think they can only write in a cabin in the woods; others think they can only write within sound of human activity. But finally all of us will write wherever we need to write to pay the bills.

As a boy I taught myself to type because I wanted to grow up to be a newspaperman—ideally, on the *New York Herald Tribune,* the paper I loved for its humanity and humor. My typing aptitude caught the attention of Colonel Monro Mc-Closkey, the commander of my army unit. The colonel had an elevated sense of personal glory, and I was a captive sergeant. He put me to work writing company histories that would exalt his feats of leadership. Once, near the Algerian town of Blida, just under the Atlas Mountains, a sirocco from the Sahara swirled through the tent where I was typing, pelting me with hot particles of sand that I sometimes think are still lodged in my scalp. The next winter, near the Italian town of Brindisi, the particles swirling through my tent were cold and felt very much like snow.

Those wartime stints at a typewriter prepared me for a life of writing in odd places, starting in 1946, when I came home and got a job on the *Herald Tribune*—my boyhood dream come true. The *Trib* building, at 230 West 41st Street, extended through the block to 40th Street, and the city room, which housed most of the editors, reporters, rewrite men, sportswriters and columnists, occupied almost the entire fifth floor.

Decades of use by people not known for fastidious habits had given the room a patina of grime. The desks were shoved against each other and were scarred with cigarette burns and mottled with the stains of coffee spilled from a thousand cardboard cups. The air was thick with smoke. In summer it was recirculated but not noticeably cooled by ancient fans with black electrical cables that dangled to the floor. There was no air-conditioning, but we would have scorned it anyway. We were *newspapermen*, conditioned to discomfort, reared on movies like *The Front Page,* in which gruff men wearing fedoras barked at each other in sentences that moved as fast as bullets. I thought it was the most beautiful place in the world.

Not everyone was as charmed by the environment. A copyreader named Mike Misselonghites, who sat along the rim of the horseshoe-shaped copy desk, arrived at his post a half-hour early every day. He would take off his coat and walk to the men's room. There he soaked and wadded up an armful of paper towels. Then he brought them back to the copy desk and scrubbed his area of the desk. Then he scrubbed his telephone and its cord. Then he lifted his chair onto the desk and scrubbed its seat and back and legs, not resting until his workplace was free enough of dirt and bacteria for him to safely go to work.

Nobody gave Mike's daily ritual a second thought, just as nobody was surprised when the absentminded music editor, Francis D. Perkins, who often smoked his pipe upside down, started a paper fire in his wastebasket. It was a community of mavericks and oddballs, held together by the common purpose of our daily voyage, equally hospitable to the portentous political columns of Walter Lippmann and the high-society gleanings of Lucius Beebe, the legendary fop, who arrived for work in midmorning after a long night of prodigious intake at the Stork Club and El Morocco, impeccably turned out in a derby, a bespoke suit, a magenta shirt and a white silk tie, his gold watch and chain suspended from a figured vest.

Much has been written about the *Herald Tribune*'s bright stars in those postwar years: the foreign editor Joseph Barnes, the foreign correspondent Homer Bigart, the city reporter Peter Kihss, the sports columnist Red Smith, the Pulitzer Prize–winning photographer Nat Fein, the music critic Virgil Thomson and many others. But the paper never forgot that its readers were an infinitely mixed stew of interests and curiosities, and it had experts squirreled away in various nooks to cater to their needs: the food critic Clementine Paddleford, the fashion columnist Eugenia Sheppard, the stamps editor, the crossword puzzle editor, the garden editor, the racing columnist Joe H. Palmer.

Palmer was typical of the paper's passion for good writing, nowhere better exemplified than in the sports section. It was in those pages, as a child baseball addict, that I found my first literary influences. The *Trib* sportswriters were my Faulkner and my Hemingway, and now I was in the same room with those bylines-come-to-life: Rud Rennie, Jesse Abramson, Al Laney. Laney, who covered golf and tennis, never took off his hat. I often paused at the sports department to watch those Olympians, wreathed in cigarette smoke, tapping out their stories with ferocious speed, especially Abramson, who seemed to have the entire history of boxing at his fingertips.

Ruling over that domain was the sports editor, Stanley Woodward. A bear of a man, built like a 250-pound fullback, he was as sensitive to good writing as a 125-pound poet. No hoopsters or pucksters played in his pages, no batsmen bounced into twin killings. Woodward had recently hired two stylists to add luster to his stable. First he plucked Red Smith from the *Philadelphia Record*, thereby presenting to a national audience the best sportswriter of his generation; in their understated humor Smith and the *Trib* were a perfect match. Then Woodward imported Joe Palmer, an English professor at a college in Kentucky, to write a column called Views of the Turf. I knew nothing about horses, but Palmer's columns, a blend of erudition and wit, strewn with allusions to Shakespeare and Chaucer, took me into a picaresque world, often straying far from "the turf." I still remember a column extolling the virtues of Kentucky jellied bourbon.

I had landed at the *Trib* on the high tide of the GI Bill of Rights, which promised to pay the tuition of every veteran who wanted to attend college. Millions did, most of them the first member of their family to go beyond high school, and much logistical chaos ensued, especially at women's colleges like Vassar, which scrambled to accommodate the male hordes. I persuaded the *Trib*'s managing editor, George Cornish, that the paper needed its own returning veteran to cover their story. That wasn't strictly true; any trained reporter could have done the job. But it got my foot in the sacred door, and for a year I wrote education articles for the Sunday paper.

In 1947 I became assistant editor of the Sunday review-of-the-week section. The Sunday editor, Robert L. Moora, was putting out the section almost single-handedly—the impecunious *Trib* had a much smaller staff than the *Times*—and he needed an all-purpose helper. Working for Bob Moora was a crash course in writing, rewriting, editing, layout and "making up" the paper in the composing room. He had been managing editor of the *Stars and Stripes* in Europe during the war and had seen every journalistic calamity. Nobody was faster at repairing a story that had gone wrong or banging out a new lead that got it started right. I loved to read those leads as he ripped them out of his typewriter, each one a miniature gem of narrative construction.

He also sent me out almost every week to write a feature article about some new or old marvel in the marvelous city. The shad were running in the Hudson. The *Queen Mary* "turned around" in a record 24 hours and five minutes. The Wrigley sign in Times Square was replaced by a 50,000-gallon-per-minute waterfall advertising Bond Clothes. The Board of Tea Examiners met for its annual sniffing of imported blacks, greens and oolongs. Ringling's circus had a new acrobat named Unus who stood on one finger. ("He's a very nervous man," his wife told me.)

I was also given copy to edit for various backwaters of the Sunday paper that nobody else wanted to touch. One was horticulture. Gardening advice articles filled two or three pages of the Sunday paper, and they were brought to me by Henry B. Aul, the editor who assembled the columns of outside experts, whose bylines are still printed on my retina: Alfred Putz, Gisela Grimm, Betty Blossom. I knew as little about gardening as they knew about syntax, and I was locked in weekly combat with their prose. Often that prose took a startling turn; there was much talk of divided bloomers, and unspeakable acts were prescribed for the maidenhair fern. Henry Aul was patient with my efforts to untangle the gnarled sentences, and he became a good friend. Every winter he took me to the flower show at Grand Central Palace, touchingly eager to make me a believer.

Not to make good friends in every cranny of the room would have been impossible. We were a tribe in motion, meeting each other at every turn: going out on stories, coming back from stories, getting clips from the morgue, reading wire copy off a machine, stretching, smoking, visiting the watercooler, stopping at each others' desks, endlessly talking shop. The talk continued downstairs at the Artist & Writers Restaurant, known as Bleeck's (pronounced Blake's), a former speakeasy with oak paneling and a mile-long bar. Located just a few steps from the paper's back entrance on 40th Street, it was a second home—and often a first home—to almost everyone who worked at the *Herald Tribune.* Reporters and editors missing in midafternoon could reliably be found there and invited back up to work.

My desk was quite near the paper's nerve center, the city desk, where the city editor, L. L. Engelking, a choleric giant from Texas forever pursuing the rainbow of perfection—like his famous predecessor, Stanley Walker, who wrote the first *Herald Tribune* stylebook—roared his displeasure at the imperfect efforts of his staff. Clustered around "Engel" were the desks of his assistant city editors and a bank of rewrite men—anonymous craftsmen revered by their peers for their ability to stitch into an enjoyable narrative the threads of information fed to them by reporters and stringers telephoning from all over the city. Phones rang incessantly.

The desks had ancient typewriters in their sunken wells, leaving only a small surface to hold the other necessities of the trade: a rotary telephone, a wire basket for copy, an ashtray, a cup of coffee, and a spike for impaling any piece of paper that a reporter might later regret throwing away. The spike was where old press releases went to die.

The rhythm of the room quickened as the afternoon turned to evening and the great reporters, like the crime reporters Walter Arm and Milton Lewis, pounded out deadline-meeting articles which, consumed over breakfast the next morning, seemed to have been as carefully composed as a story by Guy de Maupassant. Periodically the cry of "Copy!" cut through the noise as reporters summoned a copyboy to take the page they had just written. Their articles were brought in successive "takes" to the editors and were then dropped down a chute to the composing room on the fourth floor. There they were set in type and assembled with the rest of the next day's paper, and everything went down to the presses on the third and second floors to be printed and bundled and loaded onto trucks on the ground floor and rushed out into the night. None of us ever got tired of being in the same room where that daily miracle was set in motion.

One day in 1948, George Cornish called me into his office to tell me that the paper's drama editor, a kindly septuagenarian who had started his career on the *Brooklyn Eagle* in the 1890s, was to be retired and that a thorough renovation was wanted for the Sunday drama section, which included the theater, movies, music, dance, art, photography and radio. Would I be interested? I thought I already had the best job in the world; no larger aspirations had come knocking.

But I had no trouble deciding to accept. The Sunday drama section was a creature from some primordial bog of journalism, its antique typefaces and borders and drawings wholly out of step with the vibrant postwar theater of Arthur Miller and Elia Kazan and other emerging playwrights and directors. I would enjoy giving the section a fresh identity. The job would also connect with my lifelong love affair with the American musical theater, with Hollywood movies and with classical and popular music.

I moved across the city room to the drama editor's desk and found myself in a self-contained universe inhabited by some of the most skittish birds in the *Herald Tribune* aviary, including the music critic Virgil Thomson, who presided Buddha-like over a coterie of part-time junior critics like Paul Bowles; the dance critic Walter Terry, who, when he stood up to stretch, stood in fourth position; and the bibulous drama critic, who, not long after my arrival, took one drink too many and fell to his early retirement while walking down the aisle to his seat on opening night of a Broadway play. Four of the men on the drama and film staff had gone to Yale—a club within a club. I hadn't met any of them before I was dropped into their midst, an outsider who had somehow stumbled into the sacred grove. Walter Winchell reported in his column the alarming news that the T*rib* had turned over its drama section to a 26-year-old kid.

But the kid was generously received by the department and also by the industry. The Broadway theater was still a fraternity of like-minded men and women, longtime habitués of Sardi's and Shubert Alley, and Hollywood was still a company town run by a half-dozen major studios. I enjoyed working with the press agents of those two citadels of entertainment. I also enjoyed assigning articles to city desk reporters who had seldom strayed far from their beat, like Judith Crist, and who found a fresh voice writing about the arts. I gave his first byline to a skinny copyboy named Roger Kahn, who would later become a star sportswriter for the paper. In his classic book about the Brooklyn Dodgers, *The Boys of Summer*, Kahn wrote:

> William Zinsser, the young drama editor, assigned me to interview a female ice skater, pointing out that in a few weeks at Madison Square Garden the skater would play before the equivalent of a full year's attendance in a Broadway theater. Zinsser liked anomalies. Later he sent me to an off-Broadway production of *Juno and the Paycock.* "Everyone calls O'Casey the greatest playwright alive," Zinsser said, "but he can't get his stuff on Broadway." Preparing, I read five O'Casey plays in three days, which would have been a semester's worth of work at NYU if NYU admitted that O'Casey existed.

Yet in that glittering world of Broadway opening nights and Hollywood hoopla I never lost my love for the mechanics of the job. My favorite day was Friday, which I spent mostly in the composing room, making up the drama section with the printers, who also were good friends. Their union didn't allow editors to touch anything, but I could hover over the pages and figure out how to fit the section together. From a boyhood fling with a printing press I knew how to read type upside down, and Bob Moora had shown me with lightning dexterity how to calculate with a string whether a galley of type could be snaked into the jagged hole waiting for it in the page, around the ads that were already there. The composing room was a symphony of sounds, none sweeter than the irregular clacking of rows and rows of linotype machines, and no perfume was more inviting than the metallic smell that greeted me when I came down to the fourth floor, an instant link to earlier generations of newspapermen and the beautiful technology of 19th-century industrial America.

After six years as drama editor and 300-odd Sunday sections

I switched desks and became the paper's movie critic. The studios would screen their new films for the local newspaper critics, who ranged, in order of solemnity, from Bosley Crowther of the *New York Times* to Leo Mishkin of *The Racing Form*, perhaps its only provider of non-horse-related news. I spent hundreds of hours in those small and smoky screening rooms near Times Square. Then I would walk back down Broadway and write my review, trying to decipher under fluorescent light the notes I had scribbled in the dark.

My walk took me past shooting galleries and X-rated movie arcades and novelty shops, past papaya juice stands and Nedick's and Bickford's, past strip clubs and jazz clubs and cheap hotels, past Jack Dempsey's and the Latin Quarter and the Paramount, where legions of bobby-soxers once lined up on the sidewalk for a chance to swoon over Frank Sinatra, and, finally, past the horror-crazed Rialto Theater, at Seventh Avenue and 42nd Street, which beckoned me with ghoulish posters of monster movies and vampire movies, their titles dripping with blood.

From there it was only one more block to *my* block, 41st Street between Seventh and Eighth avenues, and I always felt a rush when I turned right and saw the *Herald Tribune* building rising out of an asphalt sea of parking lots and small bus terminals for buses going to towns in New Jersey that almost nobody went to. The block also had a cut-rate barbershop and a store that rented tuxedos to waiters and musicians.

I wouldn't have wanted the paper to be in any other part of town. It was where a great metropolitan daily ought to be, tuned to the streetwise cadences of Damon Runyon and the love songs of Tin Pan Alley. I was a fugitive from the expectations of my

WASP upbringing. I had left the cocoon of Princeton to enlist in the army, and when I came home I didn't go into the 100-year-old family shellac business, William Zinsser & Co., as I was meant to do, being the namesake and only son, but followed my own dream. I liked being outside the boundaries I was born into. At that time newspapermen were still a fairly disreputable social class; nobody actually *knew* any newspapermen. As a boy I had been taken with my three older sisters to Best & Co. and DePinna's, upscale stores on Fifth Avenue that expediently sold both girls' and boys' clothes, and there I was properly outfitted for a proper life. Now, in the 1950s, Best and DePinna's were still doing business, just a few blocks to the east. But in the geography of my life they were many miles away.

My desk as movie critic was so close to other desks that no nearby chitchat or phone call went unheard. Writing more than 600 movie reviews and Sunday columns at that typewriter, I learned to tune out most extraneous patter. My loudest neighbor was the society editor, Mr. Gifford, whose temper rose by the hour as he fielded phone calls from anxious mothers of the bride. At that time the daily paper ran several columns of engagement announcements and pictures of Eastern establishment girls whose nuptials, it said, would soon be held. The Sunday pages were mainly devoted to weddings, in which every bride, it seemed, wore a gown of peau de soie and an heirloom veil of alençon lace. Gifford and his lady assistant would explain their requirements to each mother, their tone of voice increasingly suggesting that she wasn't the only woman in the world whose daughter was getting married.

In the winter of 1958 an uninvited guest began looking over my shoulder. The paper was in financial distress, and one of its biggest advertisers, 20th Century Fox, informed the Reid family, who owned the *Trib*, that it might cancel its ads if its films were negatively reviewed, thereby lobbing a grenade at the sacred wall that protects a newspaper's editorial staff from its business department. Unluckily, in the next few months the studio released two of its most pretentious bombs.

The first was *South Pacific*, a movie derived from the Rodgers & Hammerstein classic that had been directed on Broadway by Joshua Logan. Logan also directed the movie, which starred Mitzi Gaynor and Rossano Brazzi. But something had gone badly wrong in almost every area of the production, including a special color process that was meant to bathe certain scenes in a gauzy Polynesian glow. I had no choice but to say what I thought, and rumblings from the Fox volcano were duly heard.

The other film was *A Farewell to Arms*, a trashy remake of Ernest Hemingway's novel, starring Rock Hudson and Jennifer Jones. As I wrote my review I had the feeling that I was also writing THE END to my career as a movie critic, and that turned out to be the case. George Cornish, the Reids' consigliere, suggested that I become an editorial writer, and I jumped at his typically diplomatic solution; he knew I would never be anyone's lackey. But I was also eager to broaden my range beyond the world of entertainment. I gladly moved into the office of the editorial board and wrote editorials for the next year, enjoying the variety of subjects and the discipline of the form.

I was replaced as movie critic by a bland city desk reporter who seldom met a picture he didn't like. Several months later I got a letter from Joshua Logan that said, "I just wanted you to know that you didn't hate *South Pacific* as much as *I* hated *South Pacific.*" Time has not upgraded my opinion of the two Fox films. Leonard Maltin's popular movie guide dismisses them both as dismal failures and gives them only 2 ½ stars.

My sudden change of jobs, reported in the press as an ethical surrender, didn't tell us *Herald Tribune* survivors anything we hadn't long known. Our much-loved house was far gone in decay. The rot had begun after World War II, when the cost of newsprint and labor rose steeply and the paper started to lose money. To reduce costs it began to reduce its coverage—and thereby also began to lose readers and advertising to the *New York Times*, its rival for the same upscale constituency.

The solution, concocted by Whitelaw "Whitey" Reid, who had become editor upon the death of his father, Ogden Reid, in 1947, was to build circulation by luring the "masses" away from the proletarian *Daily News.* He introduced a succession of cheap gimmicks—tawdry gossip columns by the likes of Hy Gardner and Billy Rose, a "personality" profile by Tex McCrary and Jinx Falkenburg (actually written by a kid press agent named William Safire), a green sports section, an "Early Bird" edition that went on sale at 8:00 P.M., a circulation-boosting contest called Tangletowns—which gradually eroded the character of his paper and drove away many of its best editors and reporters. Homer Bigart and Peter Kihss both went to the *Times.* The masses also stayed away; they knew when they were being patronized. Whitey must have resented their ingratitude.

In 1955 Whitey Reid was ousted in a palace revolution by his blustery younger brother, Ogden R. "Brown" Reid, who sometimes wore a gun and who saw Communists under every chair. Under Brown the decline accelerated; almost every month seemed to bring the departure in sadness and despair of gifted journalists who knew what they were doing and the arrival of hustlers and mountebanks who didn't.

One morning I looked up from my desk and saw a curious figure being escorted across the city room to the corner office of George Cornish. He was a small man wearing a shabby black suit and a black hat, who looked as if his job might be to stamp passports at the airport in Bogotá, and he was carrying a long black box. Word soon got around that his name was Luis Azarraga, that the box contained a secret camera that could take panoramic pictures, and that he had been hired by Brown Reid to enliven the paper with his miraculous wares.

Sure enough, over the next few weeks the *Herald Tribune*'s front page, long an ornament of American typography, was dismembered to accommodate an eight-column photograph, often running above the paper's handsome logo, that showed 40 or 50 blocks of New York skyline. There was no journalistic reason for running the pictures; they conveyed no information that the reader couldn't glean with his own two panoramic eyes. Nor were they notable examples of the art; the paper's own photographers did far more interesting work every day.

What the pictures did have was one undeniable trait: They were very wide. After a while the *Trib*'s owners realized what everybody already knew—that there isn't much demand for very wide pictures—and Luis Azarraga vanished as abruptly as he had arrived. Nobody ever did see what was in the black box.

Finally I could no longer wish away the truth I didn't want to face. The paper where I had spent so many happy years, where I had expected to stay forever, had ceased to be a happy place. When I was a kid the *Herald Tribune* had shaped the values I would take through life as a writer, and when I went to work there its older editors, especially Bill Avirett and Harry Baehr, went out of their way to help me. Now its values were lower than *my* values. I didn't want to stick around for the long illness that would finally kill the paper in 1966—a fate briefly redeemed by two late-arriving acrobats, Jimmy Breslin and Tom Wolfe, doing high-wire acts on the deck of a sinking ship.

One day in 1959 I walked across the city room one last time to the office of George Cornish. He was still sitting where he had hired me 13 years earlier—in a leather chair at the head of the long table where he presided over the daily news conference, watched by the bullet-punctured bronze bust of Adolf Hitler that Homer Bigart brought back from Berlin after the war.

A cultivated man from Demopolis, Alabama, Cornish seemed to me to be a tragic hero, in whose incremental yielding of journalistic integrity the tragedy of the *Herald Tribune* itself was played out. He never forgot that he was the servant of the owners, loyally passing along their shabbiest decisions as if he really thought they were worth a try, never betraying by the slightest tic what must have been happening to him inside, standing Canute-like on the beach as the Reids gradually washed away the foundations of his paper.

I told Mr. Cornish—I still called him Mr. Cornish—that I saw no end to the erosion and that I felt I had to resign. He said he was sorry, but I don't think he was surprised; I was only the latest in a long line of men and women who had finally lost patience with him.

I telephoned my wife, Caroline, to tell her what I had done.

"What are you going to do now?" she asked. I thought it was a fair question; by then we had a one-year-old daughter, Amy.

I said, "I guess I'm a freelance writer."

3

The Loneliness of the Freelance Writer

So began 11 years of sitting alone in a small room, the only sound the clatter of my typewriter. I thought back to my *Herald Tribune* days, when we all made regular trips to the watercooler—far more than our kidneys required; it was our village green. Now all that was gone.

What had I done? I'm a social person, dependent on people and on talk, however small. Now, overnight, I had no place to go, nobody to exchange ideas with, no paycheck or pension or support system. I would have to be my own support system, skating out on the thin ice of self-employment.

Our apartment was at 530 East 86th Street, in the Yorkville neighborhood of the Upper East Side, still heavily populated by German-Americans; Amy's babysitters were Mrs. Dodenhoff and Mrs. Mockenhaupt. It was one of those 1920s buildings in which every apartment had two maids' rooms off the kitchen. Not having a maid, we used one of those rooms for the machines that did the washing and drying that maids once did. That left one free room, and I claimed it for my office.

It didn't take long to install my writing essentials: underwood standard typewriter, green metal typing table, typing chair, a ream of yellow copy paper, a wastebasket, a box of #1 pencils, a *Webster's* dictionary, a *Roget's Thesaurus*, and my spike, which I had brought from the *Trib.* That simple object, a metal prong with art nouveau aspirations on its hexagonal green metal base, was my last tie to the newspaper furniture I had left behind.

It also didn't take me long to realize how small those Irish maids were—or, rather, how small the concerns of their employers for their comfort and needs. The room was big enough for a bed and not much more; the view across a court to the building's other wing didn't nourish the spirit. That didn't bother me as much as it would have bothered the maids—a writer who thinks he needs scenery to activate his muse is a good candidate for going broke.

Two years later our son, John, was born, and he and Amy both learned that the sound of typing would be the obbligato of their lives. My sentences almost never come out right at first, and I endlessly try to repair them. Today that hatchet work is done in silence by the delete key of a computer. But in the age of typewriters a writer would use the letter *x* to cross out uncooperative sentences and phrases, punching the key in staccato bursts of anger or frustration, as if to punish the words for refusing to fall into a graceful pattern. If Amy and John were asked to recall the most distinctive sound of their early childhood it would be the machine-gun-like rattle of xxxxxxxxxxxxxxxxxxxxxx's, again and again, behind Daddy's closed door. Daddy was not to be disturbed. It was understood that he was trying to pay the bills.

When Amy outgrew the bedroom that she and John shared and needed space of her own, she was given my maid's room. I improvised a small writing area in a loggia at the far end of the apartment, thereby removing the daily rhythms of the *x* sound from the daily rhythms of domestic life. My typewriter still roosted on its inseparable companion, the green metal typing table. That humble piece of office furniture, with its two collapsible sides that could be raised to a level position to hold papers, was—it now occurs to me—the cornerstone of clerical America well into the 20th century, used by millions of secretaries to hold the typewriter that typed the nation's letters and memos and invoices.

Then, one day, while I wasn't looking, the rest of the family was gone. Amy and John went off to nursery school and kindergarten and the lower grades, and Caroline went off to the Bank Street College of Education for a master's degree that would launch her long career as a teacher and educator. I was now entirely alone, reduced to the two activities with which freelance writers torment themselves: waiting for the phone to ring and looking in the kitchen for a snack, as if possibly someone had delivered a bag of Fritos since they last looked. I learned very early that the phone doesn't ring anywhere near as often as it should. I also learned that editors move at a pace little short of cardiac arrest. Secure in their own steady income, they are thoughtless about the professional and financial and emotional needs of their unsalaried writers. I had entered the land of the three-month response and the unreturned phone call and the check that wasn't in the mail.

The only way to survive, I decided, was not to waste energy blaming editors or trying to guess what was holding things up. They had a million excuses for not doing the job they claimed they were in business to do. Their assistant was on maternity leave. Their new publisher hadn't "come on board" yet. Their phone system was being upgraded. The accounting office had moved to Omaha. They were at the presales conference. They were at the sales conference. They were at the postsales conference. They were at the Frankfurt Book Fair. Their dog died.

One publisher signed me for a book project and then didn't send me the promised contract and advance. Weeks went by. When I finally called to complain he said that his firm had recently signed so many authors that it didn't have enough lawyers to draw up all their contracts. I told him to hire more lawyers. The contract came a few days later. Another editor initiated a project but then couldn't be reached to discuss it. When I occasionally did get him on the phone he always said the same thing: "I've been maniacally busy." I told him to hire more maniacs.

I soon hit on a standard answer for editors who whine about their own internal disarray: "That's *your* problem." It wasn't my problem. My job was to deliver their product and get on with my life. When one of my articles came back in the mail with its timeworn reply—"I'm afraid it's not quite right for us" or "It doesn't quite suit our needs at the moment"—I didn't rail at the editor for not recognizing the jewel he or she had been offered. I wrote a new cover letter to another magazine and got the article back in the mail by noon. (Mail was delivered earlier in those days.) I can still picture the mailbox at East End Avenue and 86th Street.

Two situations eased my plunge into the cold waters of freelance writing. All those movie reviews at the *Trib* had made my name familiar, and I began to get assignments to interview some rising comet like Woody Allen or to assess some new Hollywood trend. It was also the last historic moment when it was possible for a magazine writer in America to be a generalist. *Life* and the *Saturday Evening Post* were still fixtures of American culture. They needed articles every week—every *week*!—and over the next decade I wrote more than 100 pieces for them, for *Look* and for *Horizon,* a handsome hardcover magazine of culture and the arts.

But in the mid-1960s the ground under that edifice began to tremble. Television was the new behemoth, and it siphoned away the consumer advertising that was the magazines' main source of revenue. Readers no longer wanted to see a four-color ad showing the latest car or the latest kitchen appliance if they could see the car itself, hurtling along the Côte d'Azur, or the appliance itself, dicing vegetables for dinner. The big advertisements gradually evaporated, and I wondered how much longer the general magazines could last.

Then, one day, the *Saturday Evening Post* died. It had been around since the 19th century, a staple of American life, part of the glue that held the country together. Now the glue was coming loose. I filled that hole in my income with a contract to write for *Look*. Then *Look* died. Then I got a contract to write columns for *Life*. But one day, I knew, *Life* would also die; that heavyweight champ looked more emaciated every week. The game was essentially over.

The territory now belonged mainly to special-interest magazines that subdivided life into small pieces; no generalists wanted. Overnight, freelance writers became experts on hunting or fishing or guns or cars or sailboats or motorboats or motorcycles or skiing or skateboarding or brides or photography or electronics or computers or parenting or gardening or antiques or home improvement or knitting or food or wine or cats or dogs. I got in just under the descending wire. If I had been born ten years later I couldn't have survived as a generalist in the specialty world of magazines. It was the *Herald Tribune* all over again. I was in on something that was on the way out.

4

Travels with My Olivetti

To ease the solitude of writing in my wifeless and childless apartment I looked for longer projects that would get me out of the house. One was a book commissioned by the New York Public Library to celebrate the 50th anniversary of its august building at Fifth Avenue and 42nd Street. Another was a history of the Book-of-the-Month Club to mark its 40th anniversary. Both jobs immersed me for several months in a community of scholars and readers who loved their work. They also were a reminder that freelance writers should think beyond the usual definitions of the job. Many interesting stories are waiting to be told in the nonprofit world, the business world and other seemingly unlikely places.

Travel was another avenue of escape. Caroline's father's job had unexpectedly taken him and her mother to London, and in 1960 we rented a flat near their flat for three months; Amy stayed with her grandparents. I unpacked my portable Olivetti typewriter, bought some paper at the local stationer's and got ready to write whatever stories came my way. What soon came was a cable from *Life* asking me to interview the actor Peter Sellers, whose new movie, *I'm All Right, Jack* had vaulted him to worldwide fame. Until then he had been famous only in the British Empire, along with Spike Milligan and Harry Secombe, for their radio program of surrealistic antiestablishment humor, *The Goon Show*, which, once a week from 1953 to 1959, induced nine million BBC subscribers to stop whatever they were doing and tune in.

With his new movie wealth Sellers had bought a 15th-century manor house in the village of Chipperfield, 20 miles north of London, along with a mulberry tree mentioned in the Domesday Book, a flock of pigeons, 50 acres of rolling Hertfordshire and the usual assortment of potting sheds and barns. But the property hadn't turned its new owner into a manor lord of the type that *The Goon Show* often satirized. Caroline and I found Sellers in an ancient tithe barn, where previous lords had counted the livestock of their tenant farmers; he was laying 500 feet of track for his model railroad. Outside, a new Cadillac was parked under the dovecote. Sellers explained that he was a compulsive collector of cars, having bought 53 and sold 52 since his first "banger" in 1947. "They were always waiting for me when I came home," he said.

He was also a compulsive gadgeteer and an expert jazz drummer and photographer, and he had tweaked the historic house to accommodate his 20th-century addictions. On the carpet of the "music room" was a gigantic set of traps; stereo jazz blared from Chippendale cabinets, and high-tech camera equipment was everywhere. Those toys were about to be joined, he said, by a life-sized mechanical elephant, which a 77-year-old inventor had been fitfully building and which Sellers commissioned him to finish.

Later I also interviewed Spike Milligan and other former collaborators, who added pieces to the Sellers mosaic, going back to his early days doing comedy acts in nether outposts of the English vaudeville circuit, whose audiences were notoriously hostile. In Blackpool a net was strung over the orchestra pit to keep the musicians from being pelted by fruit and other missiles thrown at the actors.

I wrote my Sellers article in our flat, balancing my Olivetti uneasily on a late-Victorian table that looked as if it had been made to hold teacups, my thoughts derailed every morning by a "daily" called Harty, who came to clean and stayed to talk in a Cockney accent too clotted to disentangle. Living in the center of London, I felt all around me the presence of great writers from earlier centuries, many of them memorialized on a blue tablet attached to a house where they once lived. I was particularly glad to find Edward Lear in my neighborhood, a friend to greet on my walks. It may have been Lear's droll spirit that kept directing my attention to newspaper articles about the British love of animals—a national trait so remarkable that I couldn't resist writing my next article about it. One of its paragraphs described

> . . . an episode that occurred in the town of Plaistow when a sanitary inspector knocked at the house of Wally Farey, whose neighbors had reported some curious noises. Unfortunately for Mr. Farey, a horse answered the door. He was fined 40 shillings for "failing to abate a nuisance" and was told to evict his friend. The unfairness of the order grieved him. "To think," he said, "that a man can't take his own horse into his own home. It's not much fun living alone."

Two years later we came back to London for another stay, this time with a second child. Once again the grandparents were waiting, and so was our flat, and so was Harty, her language no less impregnable, and so was Edward Lear and the whole wonderful city. I quickly fell into its familiar urban rhythms and even had a brush with the English medical establishment. Needing an eye doctor, I was steered to a Harley Street specialist named Sir Hugh Rycroft. I was told, "He's very good. He treats the Queen's horses."

This time I had an assignment from *Horizon* to write a profile of Sotheby's, the London auction firm that had recently jumped to fame and wealth with its spectacular sales of paintings and antiques, which had become media jamborees for the titled, the moneyed and the powerful. To find out how Sotheby's had scaled the heights of that gilded world, I talked to the curators who ran its 10 departments—scholarly men and women whose knowledge extended to the smallest sliver of information about their field. I also climbed around the ancient Sotheby's building on Bond Street, which was not quite plumb, observing the tidal flow of its daily life.

That flow began every morning at a counter where people

brought objects they wanted to put up for auction. Reaching into bulky reticules, they fished out sconces and silver tea services and faience dishes in grotesque animal or vegetable form. Dowagers arrived with chauffeurs carrying rolled-up Oriental rugs, and soon the counter was piled as high as an Arab bazaar, the objects having nothing in common except that they were no longer wanted at home. It was not unlike a doctor's examination; the object was poked and prodded, and the diagnosis was often unwelcome. "It really isn't rare at all, sir, no matter what your grandmother told you. I doubt it would fetch three pounds."

Our visit happened to coincide with the imposition of huge death duties that had begun to empty the stately homes of England, disgorging hoards of art, antiques, silver and jewelry that had long been concealed from public view or weren't even known to exist. Listening to those accounts, I heard more than the crack of the auctioneer's gavel. I heard the cracking of ancient estates, the tottering of peerages and the rushing hooves of the new rich—Englishmen who had made postwar fortunes in scrap or realty, Greek shipping magnates, Italian and German industrialists, American oil tycoons and other first-generation millionaires, all hungry for the status that could be had for the raising of a hand and the writing of a check.

I also wrote a long profile of Spike Milligan, who had been on my mind since I interviewed him about Peter Sellers. As the writer of *The Goon Show* and the architect of its success, which inspired such later British lunacies as *Monty Python's Flying Circus*, *Beyond the Fringe* and the free-association plays of Harold Pinter and N. F. Simpson, he struck me as one of the most influential humorists of the postwar period. Altogether he wrote 150 *Goon Show* scripts and performed them with Sellers, Harry Secombe and some of the strangest sound effects ever put on the radio, many of them self-manufactured. What he created was an original universe of nonsense that made a certain amount of sense, eliciting cries of "Anarchy!" from the upper classes and touching the deep discontents of the other classes. Many episodes of *The Goon Show* dealt with the inanities of Her Majesty's government and the perpetual shortages of postwar England.

> "Have you seen the papers? There's a shortage of ink."
> "How do you know?"
> "They're written in pencil."

In Milligan's scripts the British people glimpsed many wistful truths about their imperial past. One show, based on some current colonial imbroglio, involved sending a 40,000-ton battleship to the Borami oasis. The plan was to break up the battleship into four-inch squares, ship it to the oasis marked DATE FERTILIZER THIS SIDE UP, and reassemble it there.

> "But the oasis is only ten feet long. They'll never get a battleship into it."
> "They could stand it up."
> "The British don't do that sort of thing."
> "Nonsense. I've seen them go to work that way."

Other shows lampooned various agencies of the crown. "We invented all these terrible ministries," Milligan told me, citing the Ministry of Explodable Bird Lime, which tried by increasingly violent means to eradicate the starlings in Trafalgar Square and eventually blew up most of London ("never mind—it'll get rid of the starlings").

A tall, blue-eyed Irishman, Milligan was given to flights of hilarity that bordered on a seizure, which cloaked a deep Celtic melancholy. Marooned in the backwash of Sellers's movie stardom, he would never regain his status as a beloved public monument, though he maintained a lifelong output of novels, children's books, comic verse and other literary forms, including a World War II trilogy about his life as a corporal contending with the military bureaucracy; the first volume is *Adolf Hitler: My Part in His Downfall.*

"Once," he told me, "my unit had halted near Tunis when an American general came along in a jeep driven by a girl with long blond hair. I later learned that it was Eisenhower. He climbed out of his jeep and came over to me. 'Corporal,' he said, 'what are those cannons?' I said, 'They're artillery, sir.' He said, 'Good, good' and rode off."

Before those two London trips my Olivetti and I had already done a lot of writing in faraway places. Twice at the *Herald Tribune* I bartered with George Cornish for an extra month of vacation, in return for which I would write articles from parts of the world that his paper was too broke to cover. Under that arrangement, in the fall of 1954, Caroline and I got married and spent two months improvising our way across colonial Africa, from the Belgian Congo to the Indian Ocean.

I had specifically promised to cover the brutal Mau Mau

revolt in Kenya, then in its third year, and I inflicted on my bride a harrowing night in a remote hotel room whose French doors almost invited the Mau Mau rebels to come out of their hideouts in the Aberdare Forest and kill us, as they had recently murdered a respected old Kenya settler. Afterward, safely back at the New Stanley Hotel in Nairobi, where Ernest Hemingway had recently medicated himself with Haig & Haig after the crash of his bush plane, I wrote—among other pieces—a five-part series on the Mau Mau war.

In 1956, under the same arrangement, we made a two-month tour of the South Seas. The seaplane that flew us from Fiji to Tahiti, the only air service then available, wouldn't be back to get us for two weeks—plenty of time to find my own tales of the South Pacific. One, hidden in a local library for 50 years, was an account of the public auction of Paul Gaugin's final effects. (Six of his "pictures" went for one franc each.) Another was about the American author James Norman Hall, coauthor of the *Mutiny on the Bounty* trilogy, whose widow, Sarah Hall, we visited at their house 10 miles outside Papeete. Hall's room and writing desk were still as he left them, his thousands of books spilling into the kitchen: tomes of British naval history, 69 volumes of the *Annual Register* of England from 1758 to 1827, 27 works by Joseph Conrad.

I wrote my Tahiti pieces in our bungalow at the hotel Les Tropiques, which had a view across the lagoon to Moorea, the twin-peaked island that inspired the song "Bali Ha'i" in *South Pacific*. But much of my attention went to the giant land crabs that scuttled past the front door and to the small animals that seemed to be chewing apart the thatched roof.

Next, in Western Samoa, we stayed for three weeks at Aggie Grey's Hotel, a ramshackle boardinghouse that was an iconic destination for every trader and beachcomber roaming the South Seas. Aggie herself, who was half-Samoan, was no less iconic—a warmhearted host to all the strays who gravitated to her front porch overlooking the busy harbor of Apia. The street that ran along the waterfront was a patchwork of interisland trading stores, copra warehouses and churches of six Christian denominations, which struck the hour at six slightly different times—an unintended symbol of the slight sectarian differences that had brought them into being.

Aggie put us up in a screened cottage in the backyard of her hotel, and there I put my Olivetti on a bamboo table and wrote more yarns for the *Herald Tribune.* One was about Robert Louis Stevenson, whose house, Vailima, was just outside town; he built it in 1890 for his extended family and spent his final years there. He had come to the South Seas in search of a benign climate for his tuberculosis and had sailed widely among the islands, which had first caught his imagination as a boy in Scotland. The Samoans called him "Tusitala," teller of tales, and when he died in 1894, at the age of 44, they carried his coffin to the top of the small mountain that rose just beyond his garden, where he wished to be buried—probably the most inaccessible grave in English letters. Or so it seemed to us as we hacked our way up an overgrown trail, pulled from the front and pushed from behind by a swarm of children who materialized from a nearby village. Stevenson's simple grave was in a clearing on the summit, inscribed with the epitaph he wrote for himself:

Under the wide and starry sky
Dig the grave and let me lie
Glad did I live and gladly die
And I laid me down with a will
This is the verse you grave for me
Here he lies where he longed to be
Home is the sailor home from the sea
And the hunter home from the hill

I wrote my Samoa articles in our bungalow at Aggie Grey's, which was raised on stilts off the ground, leaving room underneath for a transient population of dogs, chickens and children, their yelps and clucks and cries rising through the floor. It may have been the only time when external sounds and smells were woven into the fabric of what I was writing about and may even have given my sentences an extra lilt.

5

Up in the Billiard House

Another benefit of freelance life was the chance to spend summers in the country. Not that I wanted to write about the country; wherever I did my writing, it wouldn't be about the birds and the bees. But that didn't mean my family had to stay in the city. I've never let my writing define my life; I want to be a person first and then a writer. I've never worked at night or on weekends.

In 1959 we bought an aging summer house in Westhampton Beach, the nearest and plainest of the Hamptons, and began living there from June to September. At that time the Eastern tip of Long Island was one of the most beautiful small corners of America. Pale green potato fields stretched in all directions until they met the blue of water—the ocean, an inlet, a bay. The towns and villages, many with Indian names like

Quogue and Sagaponack, looked as if they had been there forever. Small family-owned stores along the main street sold the oldest necessities of life; roadside stands sold bait. Developers hadn't yet plowed up the potato fields to build immense homes for beautiful people and not-so-beautiful people of enormous wealth. The population was relatively small and stable; traffic on the roads actually moved.

The house we bought was on a street called Potunk Lane, within walking distance of Main Street. It was a shingled house that had seen many additions—and also many subtractions, some inflicted by the historic hurricane of 1938, which left a three-foot-high water mark in the parlor. The house came fully furnished, as most summer houses for sale then did. Some of its tables and dressers were known as "hurricane furniture," having floated across the bay, unclaimed, from the homes on Dune Road that were washed away by the storm.

But beneath the insults of time and weather the house still had the charm of early summer-by-the-sea American architecture and decor: bay windows, wicker settees, marble bathroom sinks, gingerbread brackets and a glider on the front porch. Its original matron was a woman named Tillie Reynolds, whose son sold it to us when he was an old man. Tillie's elaborate dinner sets, with their fish platters and soup tureens and cruets, which we never used, were part of our indoor landscape as they reposed on a sideboard in the dining room, which we also never used. We ate at a big table in the big yellow kitchen, which was the center of family life. John was in a playpen.

The house also came with a library of summer-house books—water-stained volumes, printed in bad type on yellowing paper, inscribed with the names of long-gone family members and renters and weekend guests, many of the books still clasping between their pages some scrap of topical information—the tide table for 1927 or the schedule for *Top Hat* at the Playhouse. As literature it was junk, but glorious junk, the once-famous titles still calling out across the generations: *Anne of Green Gables, Ozma of Oz, A Girl of the Limberlost, White Shadows in the South Seas, Silver Chief Dog of the North, The Mask of Fu Manchu.*

Fu Manchu! How had I traveled so far in life without meeting that devious, green-eyed wizard? Sitting under the reading lamp in a wicker chair that creaked at my slightest movement, I became the willing slave of Sax Rohmer's lush prose, impatient to learn whether Dr. Petrie and Nayland Smith would spring Rima from the inner chamber of the Great Pyramid. ("'We meet again, Sir Denis—a meeting which I observe you have not anticipated.' Those incredible green eyes beyond the globular lamp watched him unblinkingly.") On other evenings I roamed the globe in *The Royal Road to Romance*, by that tireless romantic, Richard Halliburton, whose style, I dearly hoped, wouldn't infect my own: "Higher rose the moon, fairer gleamed the Taj, now transfigured to a gleaming gossamer."

As for where I would do my own writing, that was never in doubt. Our property came with a separate building out in back—an apparition wholly unexpected. It was brown shingled, like the main house, but small and square—just one room on top of another. The second floor, reached by an outside staircase, had a balcony with a railing that ran around all four sides, somewhat like a widow's walk. It had been built by Tillie Reynolds's husband as a billiard house—and, presumably, as a refuge for him and his cigar-smoking pals after one of Tillie's decorous dinners.

The upstairs room was just large enough for a billiard table—now long gone—and for the cue-wielding gentlemen who would have stood around it. It had windows with curtains on all four sides, a rug, a fireplace, a desk, some chairs, a bed and a sink. Downstairs, there were no such civilities; it was a crude space with soapstone tubs where servants once did the family laundry. Such were the class divisions of the Jazz Age.

I installed myself and my Underwood in the billiard room. There was no phone. Except for the bed—almost certain death for a freelancer—it was a perfect writer's space, suspended above the hubbub, with a view across fields and privet hedges in every direction. Time had worn away its pretensions: The curtains were frayed, the bedspread was tattered, the desk speckled from salt air. But my *Herald Tribune* days had given me a fondness for faded glory, and I happily climbed up there every morning.

Sometimes I was called down to the house for a phone call from an editor in New York. Those calls were my lifeline to the marketplace, my fragile link to solvency. If the editor was calling from the *New York Times Magazine* I could guess what he was about to say: "We all love your piece, but three of us feel that some parts of it don't quite work, and one of our other editors has a problem with the lead, and the deputy Sunday editor thinks if you could just . . ." Then it was back up to the billiard room to fix it. That was the freelance writer's most-loathed fate: death by committee. Most of us learned to write only sparingly for the *Times Magazine*; it was just too much hassle. By contrast, I remember two editors from that freelance decade with gratitude: Don McKinney of the *Saturday Evening Post* and Dave Scherman of *Life*. Both men knew what they wanted from me, and we worked amiably together to try to achieve it. Their generosity and trust kept me going.

After a morning at the typewriter I would come down and we would pack some sandwiches and head for the ocean, never ceasing to marvel that on weekdays one of the most beautiful beaches in the world—mile after mile of pristine sand—seemed to belong only to us; the sun worshipers who owned or rented houses on Dune Road had driven back to the city on Sunday night. The ocean worked its way into our metabolism with its fierce beauty, and it would continue to nag us as a memory when we later moved to Connecticut and only had Long Island Sound to look at. Family life in Westhampton Beach was just too lonely; we needed an intact summer community where all four of us could make friends who stayed all week.

I liked to bike around the village, discovering the easy familiarities of small-town life, which I only knew from Hollywood movies. I had always wanted to grow up on a block, doing block things, and this was partial recompense. I always stopped first at Gloria Seeley's general store to buy a newspaper and to check out the latest magazines. Gloria made sure I got my *Variety* every week and always told me when the latest issue had arrived.

As the owner of a house in constant battle with the leaks and infirmities of age, I soon learned that God's angels in a summer community are the electrician, the repair man and the plumber, and I still remember them, especially Clarence E. "Russ" Rayner, the plumber, who climbed the outside stairs to my office one day to solicit my help in unclogging a sewage pipe under the backyard. He had already dug a small excavation at both ends of the afflicted section. Then, kneeling at one end and assigning me to the other, he threaded a spiky cable through the pipe and we engaged in a sort of tugging match while Russ shouted instructions across the yard ("*Push*, Bill!"). Eventually we got the blockage cleared, and I considered it a better morning's work than I had done all week at the typewriter.

Only once did I stumble on a local story too piquant to resist. I got to talking one day with a man of about 50 who was visibly down on his luck. He told me he earned some of his money at night by scavenging for lost golf balls in the water holes of local golf courses. It was a risky business, he said, because those clubs preferred to retrieve—and sell—their own golf balls. He hinted darkly at preventive measures being initiated by the clubs on outer Long Island: fences, guards, dogs. But he was determined to keep at it as long as he could.

I asked if I could write an article about his nocturnal raids. He closed up like Deep Throat. I assured him that I wouldn't use his name, but he was adamant; this was a clandestine operation, and any leak would threaten his livelihood. Finally he mentioned that perhaps if he were to be remunerated . . . A sum was agreed on—the only time I ever paid a source—and we arranged to meet at his house a few days later. He had a date that night at the National in Southampton. "They've got the best golf balls, no doubt about that," he told me. I had no doubt about that; Southampton is a renowned playground for the very rich.

Our interview took place in his garage, where he was sorting his previous night's haul into buckets according to their condition and market value. The worst balls, he explained, are called "cut-to-rubbers," hardly fit even for a driving range, and the next-worst specimen is a cluck. "That's a ball that's hacked up pretty bad." Small mountains of balls encircled him on the floor, awaiting a visit from his dealer in New York.

The garage was also cluttered with crab nets, clam rakes and poles, and I asked if those were the tools of his trade. He was affronted by the question. "All you need is a bathing suit and a bag," he said. "I hang the bag around my neck and dive in. That's about nine o'clock. The bag holds 100 balls, and when it starts to get heavy—maybe around midnight—I know it's time to go." He demonstrated the delicate detecting motion required of the feet and toes—"I can tell immediately if it's a ball or a clam or a stone"—and the rotating motion of the arms required to keep the body suspended just above the bottom of the pond. "It doesn't pay to put your feet down or you'll sink into the mud," he said. "Besides, you never know what you're going to find down there." He recalled stepping on various objects that surprised him by starting to move, including a 70-pound terrapin.

I went back and wrote my article, noting that I had changed the hunter's name to Mr. Roby "in deference to the poaching laws." It sold immediately to the *Saturday Evening Post*, where it was illustrated with appropriate squalor by the cartoonist Arnold Roth. Thirty years later, in the early 1990s, when I was moonlighting as a jazz pianist, I met Arnie Roth, who was moonlighting as a jazz saxophonist, and we've been playing gigs in New York clubs ever since. But both of us still spend our days doing what we were doing back in the 1960s—Arnie at his drawing board, I at my keyboard, meeting deadlines for editors at the other end of the phone.

So my summers in the billiard house came and went. No ghosts intruded on my thoughts—no click of billiard balls, no lingering scent of cigars. I was alone with the subject I was writing about, the noises of the world shut out. Then, one morning, I heard the strains of a carillon chiming across the field. The one sound I *can't* shut out is music; it's a language that I need to process—the contour of the melody, the structure of the chords. This carillon was very bad news. It played for a half hour and then quit for the day. But the next morning at the same hour it was back. The selections were saccharine—elevated mall music.

One of my many crotchets is that I resent people who inflict their personal tastes on people who don't necessarily share them, and this was a classic case. I did some sleuthing and found that the carillon was on the property of Basil O'Connor, about six blocks away, and that the music was a memorial to a daughter who had died. I knew the name Basil O'Connor. Who didn't? He had parlayed a friendship with Franklin D. Roosevelt, America's best-known polio victim, into the charity called the March of Dimes, a fiefdom that he ran with conspicuous self-importance. He didn't strike me as a man who would want to be crossed.

But the recitals continued; presumably they were to be a daily treat. After a week I walked manfully down to the house and looked up Basil O'Connor in the telephone book and dialed his number. My hand was shaking—I'm no good at confrontations. Maybe I had dialed the number wrong. Maybe his wife would answer. Maybe he wasn't home. He was home. After two rings the phone was answered by a deep and silky voice.

I introduced myself to Mr. O'Connor and asked if his was the carillon that I heard every morning. He said it was; he undoubtedly thought I was calling to express my gratitude. I told him that I was a writer and that many writers find music a distraction. I said I respected the fact that the concerts were a memorial to his daughter. Nevertheless—I was now beyond the point of no return—*his music was bothering me.* And perhaps it was also bothering other people in the neighborhood. Stunned silence. My listener couldn't believe what he was hearing.

Finally he found his voice and said he was only trying to bring a little happiness to the good people of Westhampton Beach. How could anyone object to that? I repeated my reasons why someone could object to that. I thanked him for his understanding, and our conversation ebbed to a halt.

The next morning, 11 o'clock came and went without music, and so did the rest of the mornings. I never heard the carillon again.

6

Writing About the Sixties

Meanwhile, in the rest of America, the '60s turned into "the Sixties," one of the most convulsive American decades. Everywhere, the ground crumbled under the existing order, especially in the areas of sexual freedom, dress, hair, language, music and respect for adult authority.

I wanted to understand what was happening, and as the old values shifted, so did my concerns as a journalist. I was one of the first magazine writers to go to San Francisco, in the winter of 1967—I was then writing for *Look*—and bring back news of the "love hippies," as they were then called, who had descended on the Haight-Ashbury district, decked out in "ecstatic dress" and drugged out on LSD—flower children running away from their parents in the slumbering suburbs. I had no way of knowing that I had climbed aboard a wave that would move at tsunami speed. By June it had washed a huge tide of ragged postulants to San Francisco, where they camped out on the streets for a so-called Summer of Love that the city's safety and health authorities hope never to see the likes of again.

I was shown around the San Francisco scene by Stewart Brand, one of Ken Kesey's original Merry Pranksters. He and Kesey and Allen Ginsberg had recently organized a "trips festival," also called a "human be-in," in Golden Gate Park that "turned on" a crowd of 10,000 people, drenching them in changing projections of liquid colors that purportedly simulated an acid trip, along with the highly amplified rock music that was the counterculture's signature sound. Brand seemed to me to be the movement's intellectual prophet. He told me he had started writing a guide to living on natural and renewable resources, to be called *The Whole Earth Catalogue.* Ultimately that book would sell almost two million copies.

Brand took me to a small industrial building with a printing shop on the ground floor. We climbed an outdoor staircase to a loft, where several young men and women were working on the second issue of a magazine called *Rolling Stone.* They didn't look as if they knew much about journalism. In fact they knew enough to create a product that an entire generation would regard as holy writ.

From there we went to a recording studio that was barely big enough to hold five rock musicians and their instruments and sound equipment. The leader was a man named Jerry Garcia. They had formed a group called the Grateful Dead and were rehearsing for their first album. The sound was louder than anything claiming to be music that I had ever encountered, and, as soon as I politely could, I ducked back out. I had heard the future and it hurt my ears.

Yet something important was happening in San Francisco. I found myself drawn into the hippies' gravitational pull as they went about colonizing the Haight, living their vision of "oneness" with man and nature, wearing the symbols of their Eastern religious longings: Tibetan mandalas and Buddhist charms and Hindu reflecting disks. One girl had a button on her embroidered Victorian gown that said I'M A HOPE FREAK—which, I thought, said it all about their movement. Many of my own long-held assumptions began to seem narrow and obtuse, well worth dumping. Soon enough their movement would lose its purity and burn itself out—too many bad trips and too many lost children. But for a flickering moment it gave America a steroid shot of possibility and joy.

When I flew home I brought to my unsuspecting family some artifacts from the new Jerusalem. One was the first album made by a group called the Jefferson Airplane. Another was a batch of Bill Graham's psychedelic posters promoting rock concerts at the Fillmore auditorium. We spread them out on the rug of our Manhattan apartment—strangers from a strange land—and were intoxicated by their swirls and colors and art nouveau lettering. Perhaps that was when the six-year-old John Zinsser first glimpsed the path that would lead to his becoming an abstract artist, in love with color boldly deployed.

All those social revolutions became my main subject in 1968 when Ralph Graves, managing editor of *Life*, signed me to write regular pieces of serious humor—pieces that would use the mechanisms of humor, such as parody and satire and lampoon, to make a serious point. Week after week I tried to hold the galloping Sixties still for 15 minutes—long enough for readers to laugh or cry over the latest cultural or military-industrial lunacy. In those days the outlandish became routine overnight, and only the humorists were pointing out that it was still outlandish. They were a mixed bag of night club comics (Mort Sahl, Lenny Bruce), cartoonists (Herblock, Jules Feiffer) and newspapermen (Art Buchwald), and all of them were dead serious.

For me there was no shortage of amazing material. One day in the travel section of the *New York Times,* tucked among the Caribbean cruises and Riviera nights, I saw an article announcing that tourists would now be able to observe nuclear tests in Nevada. Unfortunately, it said, "they will not see the blinding flashes, awesome fireballs and deafening roars of early above-ground tests," but, still, "an underground megaton shot produces a shock that can knock a man off his feet . . . and there is always the possibility of a spectacular surface cave-in and the venting of radioactive gas through unpredictable fissures in the ground." Our nuclear arsenal had turned into a fun facility. Another *Times* story, ALEUTIAN H-BOMB IS FIRED WITHOUT SETTING OFF QUAKE, announced that the Atomic Energy Commission had set off a megaton nuclear blast on the island of Amchitka without causing an earthquake after all. Such arrogant uses of military power were beyond the reach of a mere chastising editorial. They could only be hooted off the stage with ridicule.

No cultural shift caused more anxiety than the new freedom in sexual conduct and language, especially when it spilled over into the performing arts. Much pious moralizing was committed by both the defenders and the denouncers, and I didn't want to strike that tone. I was saved by the arrival on Broadway of *Oh, Calcutta!*, the nude revue produced by the British theater critic Kenneth Tynan. It touched off a rhetoric of such high self-importance that my piece "[BLIP] Is Beautiful" practically wrote itself:

> Good evening. This is David Ruskin, host of tonight's Channel 15 symposium on "The New Permissiveness: Renaissance or Ruin?" Our guests are Wendy Mankowitz, star of the new musical hit *Skin!*, in which she performs fourteen simulated sexual acts, both natural and unnatural, beating the previous record of eleven set by Cerise Hagerty in the satirical nude revue *Mace;* Ronnie Plumb, the brilliant English impresario whose "evening of civilized voyeurism," *F,* is a Broadway sellout; Ula Sjønqvist, director of the Swedish film *I Am Virgin*, hailed by New York critics for its sensitive treatment of a young girl's discovery of herself through her bizarre love for a stallion, Sven; and the Reverend Enoch Moody, whose outspoken sermon in Yankee Stadium, "Whither the Protestant Ethic?", has made him the spokesman for clergyman and laymen alike seeking a viable answer to the sexual revolt that is shaking middle-class values to their very foundation. We're in for a fun evening, so let's get started with a question to Miss Mankowitz. Wendy, I notice you're not wearing anything tonight. Is that a conscious decision? Or did you just forget to put on your clothes?
>
> MANKOWITZ: No, David, it's very definitely a conscious decision. My personal philosophy is that "bare is beautiful," and anyone who wants to wear clothes has got to be some kind of creep.
>
> PLUMB: Precisely the existential premise of my show, *F.* If audiences can go to the theater and see people doing beautiful

things to each other with no clothes on they'll go home and—you know—get in the bathtub together instead of making war. *F* is very definitely a political statement.

MANKOWITZ: Bravo, Mr. Plumb! Like I sometimes hear that people come to see me act in the nude because I just happen to have a beautiful pair of [BLIP]. What they don't understand is that I am interposing my naked body against the violence of Vietnam and Chicago and the whole military-industrial complex.

RUSKIN: Isn't that the very point you are trying to make, Mr. Sjønqvist, in *I Am Virgin*?

SJØNQVIST: Exactly! The girl in my film, Sigrid, symbolizes Truth; the stallion, Sven, represents the Establishment, and in the scene where she undresses and they [BLIP] in the meadow outside the summer palace of King Ulric, I am obviously making a deeply felt point about unilateral disarmament . . .

Corporate America also reached new heights of self-justification in that era of fat war contracts and huge cost overruns. To deal with its slippery language I invented a firm called the National Refractory & Brake Company. It made its debut in *Life* in the form of an annual report in which the chairman, Harley G. Waller, sums up with hearty pride a year that had in fact been a tissue of disasters. One was the development of a memory module to replace the crew of a passenger train:

> The module offered major savings in this age of mounting operational costs and outrageous labor demands, and we were gratified by the decision of the New Haven Railroad to test it on a commuter run from Westport to New York on September 12. It is to be regretted that the memory module apparently lost its

memory capacity and took the passengers to Providence, R.I., where they were given lunch at company expense and returned to their homes by bus. The resultant adverse publicity, while naturally disappointing, had the favorable effect of redoubling our determination to unlock the secrets of microelectronic circuitry. On balance, however, it was a year of progress, and . . .

That parody annual report was so close to the real thing, with its arcane graphs and tables and pie charts, that when it was subsequently published by Harper & Row, in the format of an actual annual report, it was mistaken for the real thing by several corporations and was filed in their reference library, under *N*. The title of that book was *ANNUAL REPROT*. An "erratum" on the inside cover explains:

The words "Annual Reprot" on the cover should read as follows: "Annual Report." The error, regrettably, was not detected until the conclusion of the press run. Your management felt that the sizable cost of reprinting the cover in this expensive process was not warranted. The saving, of course, will be passed along to you, the stockholder.

I next used the National Refractory & Brake Company on the occasion of the first Earth Day, taking as my point of departure the company's decision to build a plant for the manufacture of low-density orthoxylene refrigerants and methylated polyresins on a site along the historic Cahoga River in southeastern Ohio, adjacent to the Frank J. Fenster Memorial Bird & Wildlife Preserve, a nesting ground of the threatened godwit. My piece began with the newspaper article announcing the project, which quoted Harley G. Waller as saying, "You can be sure N.R.& B. will not be unmindful of its responsibility to minimize through the utilization of electrostatic precipitators the discharge of effluents into the Cahoga and the per-hour ratio of air-pollution emissions, notably fly-ash particulates. I have every confidence that nature and industry can live harmoniously together in this beautiful valley. After all, I'm a country boy myself."

The rest of the article was an antiphonal chorus of attacks by preservation groups, editorials ("For Shame, Mr. Waller!"), press releases, newspaper interviews, letters to the editor and interoffice memos—the whole panoply of environmental umbrage that was just beginning to seep into the national discourse. Harley G. Waller was particularly annoyed by the hostile statement released by Waldo B. Ott, president of the 21,000-member Outdoor Society, which had earlier lobbied to block the installation by United Styrenic Corp. of an underground flotation mill in Yellowstone Park for the extraction of organic sulfonates:

> Dismissing the Outdoor Society as "just a bunch of bird nuts," Harley G. Waller told reporters, "You can't turn a shovel in this country today without digging up a lot of birders and hikers. If all these nature kooks had their way, America would still be a wilderness from coast to coast. Thank God there are at least a few businessmen who care about the Gross National Product and don't get hung up on a couple of lousy godwits."

Waller's comment about bird nuts and nature kooks didn't die with that issue of *Life*. Several years later it turned up in two reference books: *Peter's Quotations: Ideas for Our Times*, by Laurence J. Peter (William Morrow & Co.), and *Whole Grains: A Book of Quotations*, edited by Art Spiegelman and Bob Schneider (Quick Fox). In Peter's book Waller is in illustrious company. Below him (on page 389) are quotations by Seneca, Abraham Lincoln, Thomas Carlyle, Woodrow Wilson, Walter Bagehot, Rabindranath Tagore, Oscar Wilde, Don Herold, Ovid, and George Bernard Shaw. Ovid would have admired such a metamorphosis.

The scent finally went cold in 1974 when I got a call from Rosy Brewer, reference librarian of the Monterey Bay Area Cooperative Library System. A local environmental writer had implored her help in locating Harley G. Waller, whom he had unsuccessfully tried to find. The writer was determined to interview Waller after seeing his quote about the bird nuts and nature kooks in Peter's book. "That quote is just so awesome!" he told Ms. Brewer. Doggedly searching a network of 450 databases, she eventually got three hits for Harley G. Waller and for National Refractory & Brake. "I was thrilled just thinking that the company was still around," she told me. But one day her librarian's intuition whispered that she was never going to find Waller, and she telephoned me to ask if that hunch was correct. She was glad to learn that it was and that she could get on with her life.

Harley G. Waller and his company made their final appearance in *Life* in a 5,000-word piece called "Upward Failure." The idea came from a real incident. I occasionally played squash with a friend at midday, at the University Club, in mid-Manhattan. Another member who played at the same time was

Frank Pace Jr., whose name I had often seen in the paper. He was one of those eastern establishment captains who alternate between positions of power in the public sector and the private sector. Earlier he had been secretary of the army; now he was CEO of General Dynamics, a giant defense contractor. My friend happened to know Pace, and whenever we met in the locker room he introduced me. "*Hel*-lo, Bill!" Pace would say, massaging me in cadences of such Southern bonhomie that we might have been old quail-hunting buddies.

One morning, reading the *New York Times* at breakfast, I saw a front-page article reporting that General Dynamics had suffered the biggest first-quarter loss ever incurred in American business. "No way Frank Pace is going to show his face around town today," I told myself. But when I got to the club that noon, there he was, all joviality ("*Hel*-lo, Bill!"). No mortification was etched on his face, no chagrin roughened his dulcet voice. He knew what I was only at that moment beginning to discover: The establishment will always take care of its bumblers. Not long afterward president Lyndon Johnson appointed Pace chairman of the nonprofit Corporation for Public Broadcasting.

I told Ralph Graves at *Life* about my locker-room encounter. I said I had never seen a man so merry after such a loss. When Pace was subsequently promoted to the public broadcasting post I called Ralph to say that it was as pure a specimen of upward failure as we were ever likely to see. Graves was elated. He seized on the phrase "upward failure" and insisted that I write a piece about it, especially when I began to report other sightings of blue-chip corporations—airlines, railroads, automakers, chemical giants—reduced to a shambles by a CEO who was thereupon squired into high government office. But I couldn't figure out how to tell the story; those companies would be well stocked with libel lawyers.

Finally the obvious solution hit me. I combined all those upward-failing CEOs into Harley G. Waller and the National Refractory & Brake Company. By then it had become a conglomerate called the American National United Allied General Corporation. My piece took the form of a special supplement to the company's annual report, saluting Waller upon his resignation to accept a high-level government post and reviewing his long career:

> Harley Waller first felt the mantle of command descend on his rugged shoulders when he was in India during World War II. The young pilot chafed under the incapacity of the C–47 to attain optimal climbing speed in flying "over the Hump" to China. He begged Lieutenant General George Stratemeyer for a chance to lead a squadron of experimental C–88s on a critical supply drop to Stilwell's beleaguered troops, pointing out that the aerodynamic properties of the C–88 were far better suited to the local altitude-mist factor.
>
> This colorful mission continues to intrigue historians of World War II. Of the 12 aircraft that roared off toward the skies of China, the sole returning plane was that of Lieutenant Waller, which he crash-landed near the base although three of its engines had dropped out. General Stratemeyer was so impressed by young Waller's feat and by his subsequent report appraising the C–88 as "not yet fully operational" that the boyish lieutenant was promoted to colonel and given command of the 92nd Bomber Brigade.
>
> Inevitably, when the war was over his record caught the eye

> of highly placed officials, who brought him to Washington for the first of several tours of public service . . . As Deputy Assistant Undersecretary of the Army in the late 1940s, Harley Waller tirelessly pressed his belief in the tank as the prime deterrent to Communist expansion in Asia. Members of congressional committees remember to this day the eloquent testimony of the engaging young man which resulted in a $200 million appropriation for production of the low-silhouette XM-42 "Grizzly Bear" battle tank at a time when advocates of air power were pleading for a buildup in fighter-plane strength.
>
> Shortly after the outbreak of hostilities in Korea, in which the new enemy MiG jet proved to have greater maneuverability than the Grizzly Bear tank and a consequent advantage in destructive capability, Mr. Waller resigned to accept what he described as "an irresistible call from private industry."

Waller's upward failures continued in a steady arc through private industry and public service. The paralysis he wreaked on America's transportation system as the first secretary of human mobility resources was indulged by President Eisenhower, who appointed him ambassador to Taiwan. Ike found him a welcome respite from dealing with John Foster Dulles. "He often tapped Harley Waller for a round of golf when the awesome burdens of the Oval Office became too heavy." Later, only the intervention of President Nixon extricated the nation from the economic calamity caused by Waller's final splurge of mergers and acquisitions. "Happily, there is one other option," Nixon said, announcing Waller's appointment as secretary of defense.

I've been told that I coined the phrase "upward failure," but that seems unlikely; the ascent of dunces to the top rung is one of the oldest laws of physics, familiar to everyone who has worked for an incompetent boss. But I enjoyed pinning down the phenomenon in a form that was both an amusement and a warning. That's what the serious humorist is in business to do, and "Upward Failure," my most complex piece of humor, is the one I look back on most fondly.

In the three decades since "Upward Failure" was published, its story has been regularly recapitulated, often with eerie fidelity, as in the case of Paul Wolfowitz, who, having steered the country into the ruinous Iraq war as deputy secretary of defense, was eased upward into the presidency of the World Bank, from which he also was subsequently ousted, leaving it in tatters, unrepentant.

7

Making a New Life

Gradually, however, the work got harder, partly because I wasn't writing in my own voice. The humorist is condemned to live by invention, constantly clothing himself in a persona that's not his own. For many practitioners of the form that's a bonus; humorists tend to be owlish folk, masters of concealment. You can read the complete works of Robert Benchley or S. J. Perelman or Ring Lardner and not learn what kind of life they led or who they really were. The famously acerbic wits of the Algonquin Round Table, like George S. Kaufman and Dorothy Parker, amused each other with their barbs over lunch, but I doubt if they were much fun over the family dinner table at night.

But I had the good luck to be born with a cheerful disposition. I have no interest in hiding; anyone who wants me can find me in the telephone book, and hundreds of readers have, bringing me rich gifts of information and friendship. I was now in my midforties, a gregarious man toiling alone in a Manhattan apartment, spiderlike, spinning words out of my innards. My *Life* pieces made people laugh, and they made people think, and they may even have done some good. But finally they had no reality. They also had no emotion. How much longer could I keep writing in a format so artificial? I needed to get closer to real people and their experiences and their feelings—and to my own.

In the spring of 1968 I got a call from the director of the Indiana University Writers' Conference, which was then one of the best of the writing workshops that bloom so profusely on American soil every summer, asking if I would teach the nonfiction writing course in July. The call activated an old dream. I had often thought I would like to teach—to give back some of the things I had learned by practicing many different kinds of journalism—and I gladly accepted. I pictured my students as young men and women not long out of college, needing only a final nudge to become the writers they yearned to be, and when a fat package arrived from Bloomington with the manuscripts of the 25 men and women enrolled in my class I pulled it open to see what their first sentences were like. They were like this:

> Parsons Junction at the turn of the century was only a village. I can still remember walking to "Doc" Wallace's Ice Cream Parlor and . . .

> Probably it was crazy for a pair of newlyweds at the start of the Depression to just climb into an old jalopy and . . .

Standing in my undershorts at the induction center, a kid from Nebraska going off to fight the Kaiser, I thought . . .

Obviously my students weren't young men and women not long out of college. Nor, as I discovered when I got to Bloomington, did they have any real talent. All of us—teachers and students of fiction, nonfiction, poetry, drama, science fiction and mystery writing—were housed in one immense building, the student union, which served every human need. On the opening day of our conference the sign outside said WELCOME WRITERS! After three days it said TRY BOWLING! Perhaps that's the real task of any writer's conference: to help the natural writers to write and the natural bowlers to find some other dream.

But I took two important lessons from my days with those older men and women. One was the power of memoir as a nonfiction genre; in my own older years it would be the form that most engaged me as a teacher. The other lesson was that I loved teaching those talentless students. I enjoyed discovering that I could find ways to help them solve their problems.

During that same year Caroline asked me what I thought about the idea of living somewhere besides New York. I hadn't thought about it at all; fourth-generation New Yorkers don't think about living somewhere besides New York. But the question gave me a jolt, and I put it to a friend, Eric Larrabee, who had uprooted his own life as an editor to start fresh in another city and another field. He said, "You know, change is a tonic." I *didn't* know that; I was afraid of change. But those four words gave me the energy to move forward.

We agreed that we wouldn't want to live in the suburbs; we weren't suburban types. We also weren't country types. We needed a community of interesting people who were within easy reach of each other—no putting on snow tires and driving through the winter night. The best possibility, we felt, was a college town where I could teach nonfiction writing. I decided to give it a shot.

I began by writing letters to college presidents and provosts and deans all over the country, explaining that I wanted to start a course in nonfiction writing. At that time the ability to write clear English about real people and events was little esteemed in American education as a basic life skill. English professors regarded nonfiction as a bumpkin at the high table of fiction, poetry and literary theory. I wanted it to be taken seriously and taught seriously, both as a craft and as a literature.

I assumed that no first-rate college would have me; I was a journalist, God forbid, with only a B.A. degree. I therefore focused on colleges that I knew to be progressive or experimental. It was a laborious chore; every letter had to be typed perfectly—no errors or corrections—and it also had to be tailored to the college's aspirations. I was holding out my hat to America by mail, not knowing who was at the other end.

Eventually the presidents and provosts and deans wrote back. Their replies were generous; nobody told me I was crazy. But as administrators they were enmeshed in the machinery of academic postponement: five-year plans, three-year plans, steering committees, curriculum reviews. Obviously this was a world where nobody was in any hurry.

The only two colleges that responded positively were newly

formed and open to fresh thinking. One was the Santa Cruz campus of the University of California, and we flew there for an interview. I had assumed that the college was in the city of Santa Cruz. But it turned out to be outside town, on a hill, in a stand of giant redwood trees. It was one of the prettiest campuses in America—but not to me. My aversion to wooded places goes back to a claustrophobic summer at a boys' camp in New Hampshire; one roadside sign I'm never glad to see is ENTERING NATIONAL FOREST. The last thing I needed was a college in a stand of giant redwoods. The other campus we visited was New College, in Sarasota, an upstart kid in the Florida system. But the college still didn't quite know what it was trying to be.

I would need to cast my net wider, and I began to write letters to all kinds of people who had some kind of college connection. Some were friends who were loyal alumni of their college. Some were professors whose books or articles I had admired. Some were just men and women who had a gift for bringing like-minded people together—God's matchmakers. Maybe one of them would send my letter to someone who would send it to someone who would hear of something. I believe that one thing leads to another. If you tell enough people about your hopes and dreams, the right somebody will eventually learn of your quest.

That somebody turned out to be R. W. B. Lewis, master of Calhoun College at Yale and a distinguished professor of American literature, best known for his biography of Edith Wharton. One evening in the winter of 1970—almost two years after I had begun writing all those letters and was fast losing hope—the phone rang in our New York apartment. It was Professor Lewis. He said he had been sent my letter by a *Time-Life* journalist named Norm Ross, whom I had never met. Norm, a Yale graduate, had come across my letter and thought Lewis might be interested. The kindness of strangers! Professor Lewis said he knew my work and would let me teach my nonfiction writing course at his college, on an experimental basis, for one semester.

On that slender thread we put our New York apartment up for sale and started looking for a house in New Haven and a school for Amy and John. Caroline, who was teaching at a public school in East Harlem, was offered a job at the pioneering Martin Luther King School in New Haven's inner city. My main employer, *Life,* was still in business, so I would continue to have that income while I looked around Yale and tried to figure out how to make a new life.

8

"Leaders in Their Generation"

My father went to Princeton, and that was the college my sisters and I grew up hearing about. He wasn't a rah-rah alumnus, but he often went to the piano after dinner and sang—along with the Schubert lieder of his German heritage—some of the songs he remembered from his college days in the early 1900s. My favorite was a football rouser that I've since sung to my own children and grandchildren, which went, in its entirety:

Wow! Wow! Wow, wow, wow!
Hear the tiger roar!
Wow! Wow! Wow, wow, wow!
Rolling up the score!
Wow! Wow! Wow, wow, wow!
Better move along

When you hear the tiger
Sing his jungle song!

But my father also sang ballads that were student favorites, sung at college parties or on the steps of Nassau Hall. His voice was warm and true, and I can still hear him singing:

I see my love on the campus, look, look.
I see my love on the campus, look, look.
I see my love on the cam-pus,
Look, you can see her now!

It was a beautiful lilting melody in three-quarter time, a perfect relic of that innocent collegiate era. My only problem was that I didn't know what a campus was, or what the girl was doing on it. But I had no trouble understanding the emotion that the song conveyed: the thrill of sighting a long-awaited love.

My father also took us to Princeton football games. Ivy League football was then a potent religion, and the Pennsylvania Railroad ran Saturday specials that deposited the faithful within walking distance of Palmer Stadium and brought them back, often marinated in alcohol, after the game. It was in Palmer Stadium that I first came face-to-face with "Yale," the archfoe I had previously known only from Princeton's great football march, which my father sang when I was barely out of the crib:

Crash through that line of blue!
And send the backs on round the end!

Fight, fight for ev'ry yard,
Princeton's honor to defend,
Rah! Rah! Rah!

Of all the Yale-Princeton games I saw, one is still vivid in my memory. In the early 1930s, under the legendary coach Fritz Crisler, Princeton's teams were a national giant, undefeated in 1933, undefeated in 1935 and, with one exception, undefeated in 1934. That's the game I remember. Sitting in Palmer Stadium on that fall afternoon, a small boy in an enveloping sea of raccoon coats and orange-and-black pennants, I could hardly wait for the referee's whistle that would send my Tigers crashing through that line of blue to smite the hated Elis, especially their cocky and charismatic end, Larry Kelley.

The game began, and no progress was made by either team. Every yard was manfully contested, and well into the second quarter the game was still a scoreless tie. Then, to my alarm, I saw Larry Kelley sprinting through the Princeton backfield. Feinting past the defenders, he broke to his left and came to a patch of open territory. He was all alone! When the pass came it was well above his head. But with maddening insouciance Kelley reached up with his right arm and tapped the ball down to his chest, where he hugged it and ran the remaining 20 yards to the goal line. The extra-point kick was good, making the score Yale 7, Princeton 0, and there it would remain. The same 11 "iron men," as they would be immortalized by the press, played the entire game for Yale.

That afternoon formed my idea of Yale—a tribe of haughty warriors, playboys who played hard—and when I grew up and went to Princeton myself, in 1940, they were still the detested rival. Somehow I never seemed to remember that my mother's family, a long line of Maine and Connecticut Yankees, was populated with Yale men, including her father, Daniel Knowlton—my own grandfather!—and her brother, Hugh. But no Yale song was ever sung in our house, no Yale lore imparted. The only Yale institution I knew was Yale Bowl.

Gradually, however, that alignment would shift. Three months into my sophomore year at Princeton the Japanese attack on Pearl Harbor propelled the United States into World War II, and my class was scattered to places with names like Guadalcanal and Anzio, which weren't in any college history book; ultimately 82 percent of us—562 men—would go into the armed forces.

Afterward I pieced together enough army credits to get my degree without going back to Princeton, and thereafter, except for an occasional reunion and an annual check, my ties to the university weakened. In that expansive postwar era it still seemed frozen in a prewar state of mind. The letters in the *Princeton Alumni Weekly* were complacent and cranky; the idea of admitting women was deplored and rejected. When I wrote to all those colleges looking for a place to teach, it never occurred to me to try Princeton. It was a pretty little island of entitlement in a pretty little suburban town, and pretty wasn't what I was after.

At Yale, meanwhile, a different narrative was being written, as I knew from writing about the Sixties and the turmoil that swept college campuses, shutting many of them down and killing four students at Kent State. A new Yale president, Kingman Brewster Jr., and a new Yale chaplain, William Sloane

Coffin Jr., emerged as national leaders and kept their university safe. Brewster also shook Yale out of its insularity as a self-replenishing haven for rich white Protestant males who had mainly gone to private schools.

"We want Yale men to be leaders in their generation," Brewster said, officially stating his mission. Unofficially, he said, "I do not intend to preside over a finishing school on Long Island Sound." He appointed a new dean of admissions, R. Inslee "Inky" Clark, who shared his vision of a student body selected on the basis of merit, not of wealth and privilege, reflecting every social and ethnic strain.

Clark's first class, which entered Yale in 1966, had more graduates of public schools than any class in Yale history. It also included far more minorities, including, most conspicuously, Jews and African Americans, and excluding—for the first time—a considerable number of alumni sons, who, until then, had been almost automatically admitted. By 1970 the Brewster-Clark revolution was complete, and those two names were spoken by old guard alumni with rage and loathing. But Brewster had yanked Yale into the future, creating an admissions process that would funnel into its classrooms and labs—especially when women were admitted in 1969—a steady infusion of future leaders.

Coffin, operating on the parallel track of moral leadership, became as reviled as Brewster for his radical activism supporting civil rights and protesting the Vietnam War—a stance that included being jailed with his fellow freedom marcher, Dr. Benjamin Spock. While other Ivy League colleges were damaged by the protests of the Sixties, Yale took new strength from the values of Brewster and Coffin, surviving a threat of severe rioting connected with the trial of the Black Panther revolutionary Bobby Seale in New Haven in the spring of 1970. I looked forward to teaching the young men and women at their college.

Not that I wasn't nervous about the move. The maxim "change is a tonic" is good for the soul but hard on the stomach, only gratifying when it's over. I worried that I wouldn't be able to find a place for myself at Yale and to fathom its 300-year-old history and covenants and codes. Just then fate delivered a break in the form of a marital calamity—luckily, not my own. The managing editor of the *Yale Alumni Magazine* ran off with the wife of the editor, and both men skipped town, leaving the magazine rudderless. Its board of directors, hearing that a real live journalist was coming to town, offered me the editor's job.

At first it seemed like a crazy thing for a middle-aged Princeton man to do. Nor was it a lofty journalistic position. But I wasn't looking for professional advancement. I wanted to create a different kind of life, and I thought, What quicker way to get to know a great university than to run its magazine? It had a circulation of more than 100,000 graduates of Yale College and its nine professional schools, which included art, architecture, divinity, drama, forestry and environmental studies, law, medicine, music and nursing. Editors are licensed to be curious, and the job would put me at the center of Yale's many interlocking worlds, its famous professors and scholars and curators and coaches only a phone call away.

I signed up for the job, and one day in the summer of 1970 we loaded ourselves and my Underwood typewriter into a car on East 86th Street and drove off into the future.

9

Old Yale and New Yale

On their third day of school in New Haven, Amy and John walked home carrying a cat. Our formerly urban children looked as if they had come from a posing session with Norman Rockwell. "It's just a loan—we don't have to keep it," they explained. I suspected that such loans to new kids in the neighborhood tend to be permanent, and that turned out to be the case; the cat, our first, had come to stay. By not returning it we took the first step in becoming proper residents of the block.

It was mostly a neighborhood of Yale families, though I seldom caught sight of the great professors who lived in our midst. Presumably they were somewhere inside the house, encircled by books and by clouds of pipe tobacco, preparing their next lecture. Their true habitat was Yale itself, which was about a mile away, and anyone who did business there—or anywhere else—had to go by car. We quickly learned that we would have to drive to the market and the drugstore and the post office and the bank and every other provider of health and welfare. That was also something new.

The orthodontist turned out to be in a little white house on Whitney Avenue. Who knew that dentists had their offices in *houses*? Dr. Gans had such perfect teeth that it was hard to look him in the eye. When I drove Amy there for her first follow-up visit he held out for my inspection a set of plaster uppers and lowers. "This is Amy," he began. "You can see she isn't interdigitating properly. She'll need accessory appliancing." I drove to the bank to ask about accessory financing, and Amy was duly ushered into that wiry world. Over the years I would occasionally see Dr. Gans scooting around town in his Porsche convertible, his smile as agleam as the chrome appliancing on its wheels. Such are the fruits of proper digitation.

So we gradually adjusted to the customs of the alien civilization we had been dropped into, as if by parachute. Caroline drove off every morning in a Volkswagen to the Martin Luther King School on Dixwell Avenue, and I settled into my own pursuits. For a home office I found an upstairs room that had a picture-postcard view of East Rock, a massive sandstone bluff that was one of New Haven's two framing landmarks, the other being its undersized twin, West Rock. Sitting there at my Underwood, I continued to send columns to *Life* until one day when my editor called and said, "Whatever you've got in your typewriter, send it to someone else." Mighty *Life* had folded, its glorious 36-year reign ingloriously over. It was a day of national mourning, but secretly I felt relief. I could finally get on with whatever I was going to be next.

My main base was at the *Yale Alumni Magazine*. Its office was just off the historic New Haven Green, where three steepled churches, a triple scoop of early Protestant piety, stood guard like deacons over the revered past. Although the Green vaguely connected me to my maternal Yankee roots, I had no idea what events made it so historic. Every spring I was startled by the long and insistent booming of cannon on the Green. As I never managed to remember, it was the annual reenactment of Powderhouse Day, the incident in which Benedict Arnold turned over the keys to the powderhouse or somebody turned over the keys to him.

As a journalistic species the alumni magazine occupies a slippery ledge in the world of academic governance. It best serves its constituents—who, being college graduates, are no dummies—when it reports the news of their college fully and fairly, operating as an independent entity with its own board of directors. It serves them worst when it becomes a house organ of the administration, extolling the president's agenda and skirting the hard truths. Many presidents have taken that route lately, replacing strong editors with compliant minions—a disservice, in the long run, to the health of their institution.

At Yale's magazine I never had any interference from Kingman Brewster, though he was then at the low ebb of his presidency. His overhaul of the admissions policy and his notorious remark, made during the ominous spring of 1970—"I am skeptical of the ability of black revolutionaries to achieve a fair trial anywhere in the United States"—had alienated the very alumni whose money he needed but whose sons were no longer necessarily wanted. Yale alumni relations were never more bitter than when I was handed its alumni magazine.

But Brewster didn't use the magazine to plead and justify. He obviously believed, as I did, that excellence is its own reward. If the magazine of a great university brings a lively curiosity to what's happening in every corner of the campus, month after month, the greatness of the university will be self-evident, its missteps forgiven or forgotten.

I took all of Yale as my domain and was dazzled by its riches. I was a generalist in a world of specialists who had carved knowledge into its smallest shards. I wanted to know what was new in every discipline, and the various deans and department chairmen obligingly let that information be extracted. Yale also seemed to have no end of collections that were unique—silver and textiles, Babylonian tablets and Tibetan sacred texts, Boswell papers and Franklin papers, literary manuscripts and music manuscripts and medical manuscripts and dinosaur bones, each one watched over by a curator or librarian in love with his or her work.

So I gradually met the men and women who gave the university its breadth and its continuity. Many older professors were generous with their friendship. One was the historian George Pierson, a descendant of Yale's founding rector, Abraham Pierson, whose statue on the Old Campus continued to gaze out at the annual crop of freshmen who roomed there. In George Pierson's long face of Yankee granite—where, to my relief, I often detected a flicker of amusement—I saw written the story of Yale reaching back to Father Abraham. That earlier Yale was a place of golden memories, a male Arcadia symbolized by the codes of Dink Stover and the songs of Cole Porter and the gridiron heroics of Albie Booth and the mixed harmonies of the Whiffenpoofs singing about the tables down at Mory's.

But as I absorbed the traditions of the old Yale I saw that the new Yale where I had landed was a different organism, galvanized by its first classes of bright women and ambitious students from urban neighborhoods who never heard of Dink Stover. In the letters columns of my magazine I watched older alumni picking their way over the emotional terrain of inevitable change, and at alumni functions I watched Kingman Brewster fending off the arrows of an anger that only slowly cooled. His own weapons were humor and charm. They were the lubricants of his leadership, enabling him to deflect an unwelcome question with a wit so urbane that nobody remembered what question had been asked.

My other base was Calhoun College, where I had been made a fellow. Yale College is divided into 12 residential colleges, each a separate principality with its own dining hall, library, social facilities, academic and cultural programs and intramural teams. Each college has a resident master, usually a tenured pillar of the faculty, who keeps the enterprise running and sets its character and tone. About 150 adult fellows also belong to each college, drawn from every corner of the faculty and administrative staff, who are encouraged to eat with the students and impart whatever wisdom they will swallow. The result is a vertical family that gives every undergraduate a sense of belonging to a small community within the large and often lonely whole.

Calhoun's master was R. W. B. Lewis, professor of English and American Studies, a much-admired man of letters. It was Lewis who had brought me to Yale, and he welcomed me into his college, groomed me in its procedures and launched me on my vocation as a teacher. Every beginner needs a mentor, and I've often thanked the lucky star that led me to mine.

One of Lewis's pleasures was to hold receptions and dinners for writers, critics and other literary lions. I recall an evening when one of his guests was Robert Penn Warren. Caroline and I, still babes in academia, listened with due deference as the wine glasses were refilled and the conversation eddied around us, eager to hear what the great poet had to say. He said a great deal, but in a soft Southern mumble that didn't seem to form itself into sentences. Afterward I asked Caroline, who was sitting nearer, what Warren had talked about. "I didn't understand a word," she said.

On another occasion Lewis arranged for Hunter S. Thompson, author of *Fear and Loathing in Las Vegas*, one of the totemic books of the 1960s drug culture, to give a public lecture at Yale. Calhoun would be the sponsor, and Thompson would stay in the college's guest suite. The problem was how to get him there; the Gonzo journalist was then living in California and was famously unreliable. Lewis deputized one of his biggest and brightest and most likable students, Mitchell Crusto, an African American from New Orleans, to borrow a car, find a brawny fellow student, meet Thompson's plane that evening at JFK airport and bring him to Calhoun College.

Perhaps nothing in Mitch's subsequent law career matched the complexity of that assignment. Thompson was so narcotized that he couldn't be pried from his seat on the plane. When that task was finally accomplished and the writer was wrestled through the airport and out to the car, Mitch gratefully set his compass for Connecticut. It was then around midnight. For Thompson, however, the plane ride had been mere prelude. He turned out to know the street coordinates of every bar in Queens, and he ordered his keepers to stop at several of them. It was almost dawn when the little band got back to Yale and the famous author was delivered to his suite at Calhoun, from which no sounds of human life were heard for much of the following day. Could he be roused in time for his lecture at eight o'clock? With considerable help, he was.

Calhoun was one of the most centrally located colleges, and that suited me well. As a city boy I was delighted to find Yale so compact, its streets flat and rectilinear. Almost the whole university was within an easy walk, including most of the graduate schools, ordinarily exiled to unloved regions of the campus. (Princeton's graduate school is at the far end of a far-off golf course.) Yale Law School, spawning ground of Supreme Court justices, was at the very center of the campus; Bill and Hillary walked unknown among us, young gods in the making. Sterling Library and Payne Whitney Gymnasium, lofty cathedrals of scholarship and sweat, beckoned to their adherents. The schools of art and architecture and drama were just down the block, the school of music was only an arpeggio away. It was a landscape of fortresses, their lineage derived from the castles and country estates of England, adorned with towers and turrets and moats and gargoyles and gates—a Gothic Disneyland. It was not like any place I had ever worked before.

10

Becoming a Teacher

I gave my writing course a plain title, "Nonfiction Workshop." I wanted to serve notice that it was a craft course and that I had no fancy aspirations; the word "postmodern" was unlikely to be heard in class, or any mention of the human condition. My aim was to teach Yale students to write clearly and warmly about the world they lived in.

The framework would be journalistic, "journalism" being defined as writing that appears in any periodic journal—as, for example, Lewis Thomas's elegant book of science essays, *Lives of a Cell*, first appeared in the *New England Journal of Medicine*, and Rachel Carson's *Silent Spring*, the book that launched the environmental movement, first ran as a series of articles in *The New Yorker*. Neither Thomas nor Carson was a "writer"; one was a cell biologist, the other an aquatic biologist. But they knew enough about writing to make complex subjects clear and enjoyable—and *useful*—to ordinary readers. That's what I wanted for my students.

My course was listed in the Yale course catalogue for the 1971 winter term. It was limited by the English department to 15 students, that being the generally regarded optimum size. Teaching writing is a hands-on task that can't be learned from a lecture in which grand truths are handed down. Those truths only get learned when a student's failure to observe them is pointed out in his or her own writing.

But a funny thing happened on the way to registration: 170 students signed up for my course. That came as an astonishment to the English Department, which was then the high temple of "deconstruction" and other faddish studies in the clinical analysis of texts. The great writers on the Yale faculty weren't the theory-obsessed English professors. They were the history professors—strong stylists like Edmund Morgan, C. Vann Woodward, Jonathan Spence, George Pierson, John Morton Blum and Gaddis Smith, who understood that their knowledge could only be handed down if they imposed on the past an act of storytelling, one that had a strong narrative pull and a robust cast of characters.

Reading the student applications for my course and interviewing the applicants, I heard a hunger for reality: "Help me to organize and express my thoughts." During the permissive Sixties their high school teachers had urged them to "let it all hang out," regardless of grammar or syntax. Now they found that they had come to college deprived of the basic tools for writing expository prose.

Making the initial cut was easy; I gave priority to seniors and juniors, whose time at Yale was running out. That still left many hard choices. I didn't want the class to be dominated by aspiring journalists: *Yale Daily News* hotshots and former editors of their high school paper. They deserved to take the course, and over the years many did. Some, like Mark Singer, Christopher Buckley and Jane Mayer, became major writers of articles and books. Others became influential editors: John S. Rosenberg, editor of *Harvard* magazine; Roger Cohn, editor of *Audubon* and *Mother Jones*; Kit Rachlis, editor of *Los Angeles* magazine; David Sleeper, founder of *Vermont* magazine; Kevin McKean, editorial director of *Consumer Reports*; Dan Denton, founder of several magazines in the Sarasota area; Janice Kaplan, editor of *Parade*; and Corby Kummer, senior editor of the *Atlantic* and a respected food writer. I didn't teach him anything about food—one reason for his success.

But I also wanted generalists—men and women majoring in a broad range of arts and sciences; I was looking for the next Oliver Sacks as much as the next Gay Talese. I accepted one senior history major, Lawrie Mifflin, because I was struck by her interest in sports. As a member of Yale's first contingent of women, she had been an activist for the formation of women's teams—an idea that the administration hadn't leaped to embrace. ("*Field* hockey? At *Yale*?") I felt that sports was rich terrain for nonfiction writers; some of the country's most intractable social problems were played out there: women's rights, drugs, steroids, racism, violence, betting, huge television contracts, the financial seduction of college athletes, and many more. I wanted those issues to be aired in the class.

As it turned out, Lawrie Mifflin would make history of her own, eventually becoming the first woman sportswriter on the *New York Daily News*. She covered the New York Rangers for eight seasons, first for the *News* and then for the *New York Times*, where she later was deputy sports editor for five years. She also covered the New York Cosmos during the Pelé years and at various Olympic games she became an expert on gymnastics, diving and horse-show jumping.

Another chance discovery was a blue-eyed Irish kid named John Tierney, whom I met one night in 1972 at a student social hour. Calhoun's freshmen had been exiled to a remote annex during a renovation of the Old Campus, and fellows were encouraged to drop in and make them feel less forsaken. I got to talking with Tierney, who told me he had come to Yale to major in mathematics. But as he talked I detected a most unmathematical vein of humor. He asked what I was doing at Yale, and he said he thought that would be interesting work. Could he take my course? Maybe later, I said; he was only a freshman.

But when the next term came around I couldn't resist letting him in. The writing he did was fresh and he had a bent for science. After graduating from Yale he would become a freelance science reporter for *Esquire*, *Newsweek* and *Rolling Stone* and would write humor pieces for the *Atlantic*, *Playboy*, *Spy* and *Outside*. In 1990 he was hired by the *New York Times* as a general-assignment reporter and later became a columnist on its op-ed page. One day in the 1990s I met him in New York at an antiques show with his parents, who were visiting from Pittsburgh. Hearing my name, his mother, a longtime schoolteacher, threw her arms around me in a hug of maternal gratitude. I had saved her son from being a mathematician.

It was the generalists who gave the class its breadth. Although they weren't journalism bound, they were eager to learn to write well for whatever career they might pursue. One woman student, Perry Howze, would find time among other jobs to co-write the movie *Mystic Pizza*. A rock musician, Gary Lucas, said he was proud of the "discursive style and rhetorical flourishes" that had won him a writing award in high school. I showed him how to get rid of those award-winning elements and urged him to try writing about rock music. He did, and he immediately began to sell rock reviews and articles to the *Village Voice* and various music magazines. Many years later, in New York, between European tours, he would call and invite me to one of his gigs in a downtown club. The club was never easy to find, carved out of some pitch-dark Greenwich Village cellar, nor was Gary, clad in black and enveloped in the blackness of the room. But when he played his guitar he was a man totally fulfilled in his chosen work.

A law-minded student, Roanne L. Mann, would become a federal judge at the United States District Court in Brooklyn. Asked to recall my class, she said, "My work as a judge requires that I communicate clearly in my written opinions. I cannot prove a direct connection between my judicial style and an undergraduate journalism course I took many years ago. Nevertheless, Bill Zinsser's class was one of the highlights of my years at Yale, and, as we say in the trade, one may reasonably infer that it had its intended effect."

The class met in a small room in Calhoun College. All the residential colleges had seminar rooms somewhere in their Gothic innards, many of them architecturally surprising in their homage to some long-vanished English ideal. The seminar room thus became the latest of the spaces where I would do my work. I never did any writing in those rooms, but I did a lot of thinking about how writing gets learned and taught and nourished.

I don't recall that I brought to the course any pedagogical scheme. I would teach mainly out of my own experience; what had worked for me as a writer would probably work for my students. What I would teach would be good English—not good journalism, or good science English, or good sports English, or any other kind of English. I would teach the plain declarative sentence and the active Anglo-Saxon verb. Passive verbs would be discouraged; so would Latinate nouns like "implementation." Clarity would be the main prize, along with simplicity and brevity: short words and short sentences. My favorite stylists would be invoked: the King James Bible, Abraham Lincoln, Henry David Thoreau, E. B. White, Red Smith.

On those plain precepts my little craft set sail. Every week I assigned a paper on one of the forms that nonfiction commonly takes: the interview, the technical or scientific or medical article, the business article, the sports article, the humor piece, the critical review, writing about a place. I would explain the special requirements of the genre, often reading one of my own pieces to demonstrate how I had tried to solve the problem, or reading passages by writers I admired who had brought distinction to a particular form: Alan Moorehead, Joan Didion,

V. S. Pritchett, Norman Mailer, Garry Wills, Virgil Thomson. I wanted my students to know that nonfiction has an honorable literature; they were entering the land of H. L. Mencken and George Orwell and Joseph Mitchell.

Mitchell had been the most influential journalist for nonfiction writers of my generation. His long *New Yorker* articles about the New York waterfront were gems of reporting and humanity; the "ordinary" people he wrote about were never patronized or judged. But he had perversely allowed his books to go out of print, and the students in my class had never heard of him until I brought in some passages to read. One of those young men, Mark Singer, would grow up to be Mitchell's heir in his own generation; his *New Yorker* portraits of assorted rogues and brigands and mountebanks make their point with dry amusement, not with censure. Several years after Mitchell died in 1996, at the age of 91, Singer wrote a commemorative piece in *The New Yorker* that mentioned where he first heard about him. I like to think that in some seminar room at Yale today there's a student who will grow up to be the next Mark Singer.

When I first taught my course I assumed that I would achieve most of my teaching with my tendentious little talk explaining the form that the students had been assigned next. I sent them forth to do a travel piece, or a sports piece, or an interview in full confidence that they would apply all the hard-won principles I had so lucidly imparted. But when their papers came back, only about 20 percent of those principles had made it onto the page; pitfalls I had specifically warned against were repeatedly fallen into. The moral was clear: Crafts don't get learned by listening. If you want to be an auto mechanic you take an engine apart and reassemble it, and the teacher points out that you have put the carburetor in wrong. I would need to get my hands dirty making sure every carburetor was properly installed.

After that I began every class by reading aloud good and bad examples from the student papers of the previous week. Perpetrators of bad examples were never identified; the rest were named and praised. Writers, I learned, are one of nature's most unconfident species, in constant need of assurance that they are not doomed souls. After class I handed back the students' papers with my corrections and comments and encouragements. That's where the real work got done.

The overwhelming sin was clutter. It was in that Yale class that I became a fierce enemy of every word or phrase or sentence or paragraph in a piece of writing that wasn't doing necessary work. To this day, what my students most remember was my pruning of the weeds that were smothering what they wanted to say. One of those students, Katie Leishman, who would become a prolific writer of medical articles, recalled many years later that

> Bill Zinsser's process went beyond editing. What demanded removal was often material that a competent editor would leave untouched. The stuff wasn't always badly written; it often *sounded* great. The hitch was, it wasn't true.
>
> Not that it was factually inaccurate. It just wasn't genuine for a particular student. Somehow Bill was able to coax us toward that self-recognition. Few editors can. Besides, it's something a writer ultimately has to realize alone.
>
> Today I still wonder why you can never internalize the exercise, why you can't stop yourself before the nonsense is on paper. You have to see it to reject it. It's like an immune

response: if it doesn't feel like you, it has to go. In Bill's vision, once the clutter (and the baloney) are gone, the writer emerges and the work acquires its force. Anything—from African violets to nuclear physics—can be explained to a reasonably thoughtful reader. Anything can be made interesting.

Bill showed us that good writers are inimitable, and why. It is the choice of language, of course, but it is also the use of time. He introduced us to writers who wrote an aphorism a day, and others who had a Sunday newspaper column, and still others who produced an article every five years. Writers pace themselves differently and are drawn to different subjects accordingly. That connection, he taught us, should be honored.

The Yale English Department, acting with a speed wholly uncharacteristic of college English departments, saw what was happening and jumped aboard the train. As a stopgap it hired several New York editors to come to New Haven and teach courses that roughly replicated mine. Then it went about establishing its own strong program of expository writing. What all of us learned was that organizing and writing a nonfiction paper is largely untaught in American schools. Yale responded promptly to that dismal news, and its commitment to nonfiction writing, including a writing tutor in every residential college, has been in place ever since.

11

A Summons from the King

In the late spring of 1973 I got a call from Kingman Brewster's office saying that the president wanted to see me. I couldn't imagine why. He and I had only met casually at Yale events and never had a conversation.

An appointment was made for the next day, and I walked over to Woodbridge Hall, the small limestone building that belies with its modest scale the power that emanates from the second-floor chamber of the king. Brewster was seated at an antique English desk, looking, as always, presidential; wherever he went on campus, there was no doubt who was the boss. Now he got up and greeted me warmly, showed me to a chair, and came right to the point: Would I accept the position of master of Branford College?

The question caught me by surprise. I knew that the terms of several masters were ending in June; during the winter Caroline and I had been interviewed by a search committee of Branford students. But I had long put that interview out of my mind; Yale's college masters were tribal elders, professors with sterling academic credentials. (Some masters during that period were the sociologist Kai Erikson, the future Yale president A. Bartlett Giamatti and the art historian Vincent Scully.) I was a layman from outside the gates. Probably my main credential was that I made myself available to students, my door at the alumni magazine open to anyone who climbed the stairs.

I didn't know anything about Branford College and hardly knew where it was. I asked Brewster how long a term he had in mind. He said five years. That was a long time to sign up my family for a life that was peculiarly abnormal, holed up in a Gothic fortress with 400 post-adolescents who would always, in one way or another, be at the door. The offer appealed to me, but I had one big reservation. Just that week Caroline had been selected as head of a New Haven school, St. Thomas's Day School, which was looking for strong new leadership. That job would leave little time or energy to be the helpmeet of a college master—a role that had long been played with genteel subservience by masters' wives, who poured a million cups of tea, presided over dinners and other social functions in the master's house, and kept a maternal eye on the students and their activities and their dates from Smith and Vassar who descended for football weekends. Some wives were even more visible than their husbands. Alison Henning, the spirited wife of the master of Saybrook College, reigned, it seemed, almost as long as the Queen of England, and with more panache.

But now it was 1973 and that division of labor no longer seemed appropriate. Feminism was gaining momentum and wives were beginning to have important separate careers. I wouldn't expect Caroline to be a stay-at-home master's wife; I would want her to be a role model for Branford's women, leaving the college every morning to go to her own work. I told the president I didn't want to impinge on her new job and should probably decline.

Brewster didn't waste any time tugging at that Gordian knot. "Caroline's job comes first," he said. "Whatever she does for Branford would come after that." The master's house, he explained, came with a daily housekeeper and was maintained by Yale's physical plant. He told me to take whatever time I needed to discuss the matter with my family and perhaps with some other college masters.

I asked him to tell me about Branford and its culture. All I knew was that it had a liberal reputation. During the threatened Black Panther riots in the spring of 1970 its master, a biology professor who still saw himself as a child of the Sixties, played a reckless political game that put the college at risk and left a residue of anger. Brewster told me that he was no longer on speaking terms with the master and that he dealt solely through the dean. That didn't sound so good; obviously there was a lot of healing to be done. I asked him who the dean was. He said it was Carlos Hortas, a junior professor of Spanish. Then he said, "You'll like him." That was all I needed to know. I told Brewster I would give him my answer soon.

That night at our dinner table I reported the news from Woodbridge Hall, explaining that it would mean a radical upheaval in our family situation. Nobody seemed to think that was any big deal; Amy and John, who were then 15 and 12, had always trusted us to try to give them an interesting life, and Caroline's main concern had been addressed by Brewster. Our instinct was to go for it—pending a tour of the master's house that would be our new home. We had no idea what it would be like.

Actually none of us had quite got the hang of living in the house we were living in now. Busy with our various lives, we hadn't given it the attention it deserved; the yard cried out for love. Our neighbor happened to have one of the biggest oak trees in Connecticut, whose branches, hanging over our backyard, dropped more leaves every fall than I had ever imagined owning or having to remove. On winter mornings the sound of ice being scraped off car windshields was a prelude to starting the day—and, with luck, the car. If we moved into Branford College we would be moving into the city, its shops and other mechanisms of life support just a block away.

During the next week I met with some present and former masters to see if the job had any rude surprises I should know about. They warned me that it had unique strains as well as unique satisfactions, but they all felt it was worth doing and urged me to accept.

I also took a walk—my first—around Branford College. Like most adults connected with Yale, I only knew the college where I was a fellow and had almost never visited the others. Now, standing on High Street, across from the historic Old Campus and its statue of Nathan Hale, I was struck by Branford's size and centrality. It was at the heart of Yale, literally and metaphorically; Harkness Tower, Yale's signature icon, rose next to the main gate, a Gothic arch tall and wide enough to accommodate Robin Hood and a dozen of his galloping horsemen. The college was surrounded on all sides by a moat, about six feet wide and basement deep. Even without water it was a formidable barrier.

Entering the gate, I found myself in a spacious grass courtyard that was the centerpiece of an architectural jewel, Harkness Memorial Quadrangle, which had been designed by James Gamble Rogers and built after World War I, presumably by an army of master stonemasons, as a unified network of student rooms. The stone was a pale pink granite, and every component big and small—towers and ramparts, niches and gargoyles—made a harmonious whole. The windows were casement windows, many with small figures frolicking in the windowpanes.

When the residential college system was created in the early 1930s, that quadrangle was partitioned to form Branford and Saybrook Colleges. Saybrook got the taller eastern half, which had less sunlight and a smaller main court. Branford's half was all sun and air, with a large courtyard, lower buildings and three small courtyards of unusual charm, each one different, named Brothers in Unity, Linonia and Calliope. The soaring Branford dining hall, at the north end, had been formed by gutting an entire wing of student rooms. So had Saybrook's dining hall, and the gates between the two colleges were locked to preserve their separate identities. Branford was the college on all the postcards and walking tours.

But where was the master's house? Almost all the other colleges, built from scratch, had masters' houses with the standard appurtenances: front door, roof, dormers. No such domestic shapes caught my eye as I stood in Branford court. I asked a student where the master's house was, and she pointed to the south wall of the quadrangle. The first two floors, she said, were the master's house; the upper three contained the college library, the dean's suite, the guest suite, and various fellows' suites and offices. It didn't look like a house to me.

A few days later we were given a tour by the master's wife, a Frenchwoman whose taste ran to red velvet. The house was deeply longitudinal. "It's OK," Amy once told a friend, "if you don't mind living inside a castle wall." The first floor was public space, used for college occasions, including a living room and a formal dining room with a long table. Beyond the dining room was a pantry lined with cupboards containing Branford-crested dinner sets and a stainless steel table for assembling the master's formal dinners, and beyond that was a kitchen with a large industrial stove. Together they made a diorama of the college system in its earlier years, when masters ruled with seigneurial aplomb, entertaining their favored fellows and friends and not having much truck with the students. I didn't picture our family spending much time at that long table. Our dining room would be a small alcove off the kitchen that had a table just right for a family of four.

The second floor, which was somewhat like a railroad flat, bravely tried to conceal its dormitory origins. Former student rooms had been reconfigured to create a family living room, a master bedroom, some large and small bedrooms of indeterminate parentage and a multitude of bathrooms. Just as it didn't look like a house, it didn't feel like a house, but we could make it work, and I called President Brewster to let him know.

The house needed a lot of remedial labor. Incoming masters' families were allowed to refurnish and redecorate the house in any style of their choice. Our main decision was to have the whole place painted white. We also decided to use our own possessions wherever we could, even in the public rooms: paintings, prints, books, Steinway piano, Chinese scrolls, Balinese textiles, African carvings and other objects brought home from our travels. We wanted the students to know our interests and tastes and to feel that a real family lived there.

On July 1 the former master moved out and the cleaners and plasterers and painters and plumbers and electricians moved in. We sold our house next to the giant oak tree and waited for Yale's house to be ready and the fall term to begin.

12

College Master

One of the shocks awaiting us when we moved into Branford was that Harkness Tower, which rose almost directly over the master's house, had a 44-bell carillon. The dreaded instrument, resentfully remembered from Westhampton Beach, was back in our life, and this time there would be no chasing it away.

Probably I had heard the carillon from a distance as I walked across the campus and thought of it as mere perfume in the academic air. But that was from a distance. Now, overhead, the giant bells were less euphonious, their tone cloudy and not quite musical. They were also very loud. Nor was there much of a repertory for those bells. Occasionally one of the student carillonneurs, striving for relevance, would play a Beatles song, but I don't think John Lennon would have taken it as a favor.

Over the years I would have some success in reducing the number of recitals inflicted on us inmates down below—a quixotic campaign that landed me in the comic strip of a student newspaper. Master goes berserk.

My office was on the ground floor of the house, just inside the main gate, and that's where my Underwood typewriter and its green metal table found their next resting place. Sitting next to the window, I could watch the students walking to and from the dining hall or playing Frisbee on the grass. They could also watch me as I worked. So could all the people being led on a walking tour; outside my window was where the guides brought their tour to a halt. Pointing upward to Harkness Tower, they would say, "We're now under the tallest freestanding structure in the Western world." It wasn't a comforting thought.

Music was never far away. We had moved in with a generation of young people who liked their music loud and who brought to college enough electronic equipment to guarantee that outcome. In warm weather they left their windows open and treated the whole quadrangle to their rock favorites. My preferred musicians were the black students who sat in the court slapping out intricate rhythms on African and Caribbean drums.

So we settled in to our new habitat and adjusted to its quirks, including Felix, the last descendant of the cat that was brought home from school. Felix was impatient to explore the yard, but it wasn't obvious—even to a cat—how to get out of the house. He ended up rooming with John, whose bedroom was over the kitchen. Outside its window, six feet away, were three Gothic arches connected across the top by a stone ledge exactly as high as John's windowsill. Night after night, year after year, Felix made that acrobatic round-trip, leaping to the ledge, scrambling down one of the columns to the courtyard, returning later to climb up and then jumping back into the window.

Amy and John made good friends among the Yale students. Amy's closest friends were a group of sophomore women—bright torchbearers of Yale's new female population. Another friend was a tall African American senior named George Lewis, a superb trombonist who would go on to a fine career as a musician. Many years later I asked Amy how she happened to strike up that friendship. She said, "I only remember that he was always very kind to me."

John's friends were the loners and apparent losers he had met—an adopted kid brother—while playing endless games of pinball in the college basement and watching endless episodes of *Star Trek* in a small common room that seemed to have no other use. Long afterward he told us that he had walked every inch of Yale's infamous steam tunnels. He brought into our lives a steady trickle of young men with an offbeat vision whom I might otherwise have missed. All of us were caught up in the domestic rhythms of the 400 young people living in our midst, never surprised by a knock on the door at some odd hour.

Many knocks could be expected just before the students went home for Christmas vacation. They came bearing their houseplants for us to keep warm and watered, often accompanied by a note specifying the care and feeding that their plant absolutely couldn't live without. Caroline put all the plants on the tile floor of one of our unused bathrooms, which began to look like a miniature rain forest. During the three-week vacation she watered them on a schedule of her own, ignoring the sacred instructions, and they all thrived. One year a student left a plant that was a smaller specimen of one that she herself had raised to a larger size. When the student came back she couldn't resist giving him the bigger one, pointing out how well it had done. He was amazed and only slowly figured out what growth hormone had been at work.

Other knockers at the door might ask if they could borrow some props for a play—period furnishings from a period household. One year a silver cocktail shaker, a long-forgotten wedding present, was needed for a Noel Coward play. That was the last we saw of it, except in the play ("there's our cocktail shaker!"). It was the dawn of the age of entitlement, and we were seeing its first student generation—young people so focused on their own needs that it didn't occur to them to think about anyone else's.

On a typical day I would spend the morning at the *Yale Alumni Magazine*, working on the next issue. At midday I came back to the Branford dining hall, pushing my tray through the line and sitting at one of the long tables with whatever mixture of students and fellows was already there. Different college masters had different styles; there was no rulebook for the job. Some of them loved administrative detail and kept formal office hours. My style was ad hoc; I got much of my business done by talking to students in the dining hall or in the courtyard or on the street, or maybe at the pinball machine.

"Can I make an appointment?" they would ask. "It's kind of a rush."

"How about right now?"

"Well, there's a play by Harold Pinter that some of us want

to put on in the dining hall and we wondered . . ." That was office hours.

In the afternoon I would work in my Branford office on college matters. I might meet with students on the committee that originated the 12 residential college seminars that the college offered every year, for full Yale credit. Or I might work on my own course—correcting student papers, meeting with students or teaching my class in one of Branford's seminar rooms, one of them reached—surprisingly to the students—by climbing a spiral stone staircase inside Harkness Tower.

Around six, I would walk through the dining hall—a final check for questions and concerns—and then go back to the master's house for our approximation of a normal American family evening, often heading out again after dinner for a meeting or a concert or a talk by some visiting personage. There was no shortage of enticements; the campus was awash in posters. One evening I attended a reception in Timothy Dwight College for an aspiring politician. I found myself standing next to a slender man who didn't seem to have anyone to talk to. He said his name was Jimmy Carter and he was running for president of the United States.

The machinery of the college was run by the master and the dean and their immensely capable administrative assistants, LaRue Brion and Liz Sader, who *really* kept the place going. The dean dealt with the students' lives at their most immediate: academic performance, discipline, physical and emotional health. Carlos Hortas, as promised by Brewster, was easy to like, a man committed to the young men and women entrusted to his counsel. Their stress levels rose as the term progressed and their course work didn't. Sometimes Caroline and I would be jarred awake at 3:00 A.M. by a piercing yell from the courtyard: "Does anybody ca-a-a-re?"

For such lost souls Carlos had the power to dispense a potion as avidly sought as a papal indulgence: the dean's excuse. It granted a student more time—*more time!*—when a paper or an exam was due or overdue. Carlos once showed me a batch of supplicating notes from students, many of them slipped under his apartment door well after midnight:

> Dear Carlos: I desperately need a dean's excuse for my chem midterm which will begin in about one hour. All I can say is that I totally blew it this week. I've fallen incredibly, inconceivably behind.

> Carlos: Help! I'm anxious to hear from you. I'll be in my room and won't leave it until I hear from you. Tomorrow is the last day for . . .

> Carlos: I left town because I started bugging out again. I stayed up all night to finish a take-home make-up exam and am typing it to hand in on the tenth. It was due on the fifth. P.S. I'm going to the dentist. Pain is pretty bad.

> Carlos: Probably by Friday I'll be able to get back to my studies. Right now I'm going to take a long walk. This whole thing has taken a lot out of me.

> Carlos: I'm really up the proverbial creek. The problem is I really *bombed* the history final. Since I need that course for my major I . . .

Carlos: Here follows a tale of woe. I went home this weekend, had to help my Mom, and caught a fever so didn't have much time to study. My professor . . .

Carlos: Aargh! Trouble. Nothing original but everything's piling up at once. To be brief, my job interview . . .

Hey Carlos, good news! I've got mononucleosis.

If the students went to Carlos to ask how to get through tomorrow, they came to me to ask how to get through their lives. They were workaholics obsessed with getting high grades. I saw four kinds of pressure operating on them:

One was parental pressure: the pressure to succeed.

One was financial pressure: the need to pay back all those loans.

The third was peer pressure: the belief that everybody else was getting better grades.

The fourth was self-imposed pressure: the need to follow the certified road to riches. No detours or risks.

Mainly I told them that there is no "right" way to succeed. I told them that the road ahead would have more turns than they imagined—plenty of time to change jobs and careers and whole ways of thinking. What I wished for all my students was a release from the clammy grip of the future. I told them to enjoy the day and its friendships and its unscheduled pleasures. I told them they would learn more by going to the new Woody Allen movie on Saturday night than by studying until midnight in the library.

I kept a special eye out for students who were preparing for careers in fields like medicine or law because their parents wanted them to, though they themselves harbored a different dream. I urged them not to follow expectations that weren't the right ones for them, citing my own career as proof that an unorthodox life was possible. I reminded them that they had only one life and that it was theirs to live. I told them not to be afraid to fail—that failure is a better teacher than success.

A senior named Josh Lobel, on his graduation day in 1976, left a note for me that said, "You have constantly stood up for the strangling student—trying to make Branford a place where tensions and pressures will be eased."

My other constituency was Branford's corps of 150 fellows and associate fellows. The fellows were professors and scholars from every branch of the university. Many were giants in their field: the psychiatrist Albert J. Solnit, director of the Yale Child Study Center; the historian John Morton Blum; the Estonian poet Aleksis Rannit, now curator of Yale's Slavic and Eastern European collections; the physicist D. Allan Bromley, director of Yale's nuclear accelerator; the French horn player Willie Ruff, professor of music and curator of Yale's Duke Ellington Fellowships; the Nobel Prize–winning cell biologist George Palade; the professor of public health George A. Silver; the dean of the Yale Drama School Robert Brustein.

One fellow who lived in the college was John Rodgers, an eminent geologist who collected mountain ranges and paid a formal call on me every June to notify me that he was off for the summer ("this year I'm doing the Urals"). He had a Steinway grand in his suite and once asked me if he could give a concert in the dining hall that would illustrate the successive stages of Scriabin's growth as a composer. I had no trouble agreeing; the recital was a treat and an education.

Quite a few fellows were from the School of Medicine. They were doctors of unusual humanism, who took a lively part in the life of the college. One, the neurologist Gilbert H. Glaser, was doing research on epilepsy that interested Willie Ruff, who also lived in the college as a resident fellow. Ruff taught a course on rhythm at the School of Music and wanted to know more about its sources than he had learned as a member of the longtime Mitchell-Ruff jazz duo. He proposed that Branford College originate and cosponsor, with the School of Medicine and the School of Music, a three-day symposium at which experts from various disciplines would explain "how rhythm is at work in all they do."

Ruff was a fountain of intellectual ideas in the life of the college, and I was always a willing coconspirator. For his symposium on rhythm he rounded up many unusual suspects. The metallurgist Cyril Smith talked about symmetry and dissymmetry in molecular structure; the geologist John Rodgers talked about the strict numerical repetition in the formation of crystals; the zoologist G. Evelyn Hutchinson talked about the rhythmic movement of lake waters; the art historian Malcolm Cormack talked about rhythm in painting; and Gilbert Glaser explained the role of the brain in the physiology of rhythm.

For three days the rhythmologists—who also included the harpsichordist and Bach scholar Ralph Kirkpatrick, the poet John Hollander, the dancer Geoffrey Holder and some Javanese gamelan players—turned each other on. To me, it was a classic example of Yale's college fellowships as a horizontal community, where scholars from various disciplines could learn from each other. Most universities are vertical organisms, each department keeping to itself and keeping its knowledge in a separate box.

My friendship with Willie Ruff would become a rich bonus of my Branford years. In 1981 I traveled to Shanghai with him and his pianist partner Dwike Mitchell to write a piece for *The New Yorker* about their concert introducing live jazz to China, and in 1984 I wrote a book about the lives of the two men as musicians and teachers, called *Mitchell & Ruff*, which is still in print and is regularly used in schools.

Branford's fellows met twice a month. Their ranks also included a category called associate fellows, who were mostly local businessmen and civic leaders and Yale alumni with some connection to the college. The first meeting of the month was purely social—drinks in the master's house—and Caroline and I enjoyed greeting all those fellows and their spouses, familiar faces at the door. Playing host to such a constellation of scholars, each one a custodian of specialized knowledge, I took pleasure in being a generalist; in my writing or my editing or my travels I had touched some small corner of almost every discipline the fellows brought into the house. Generalists, I found, can move mountains, or at least foothills.

The other fellows' meeting, on the third Thursday, was a so-called business meeting, held in the Fellows' Lounge after dinner, at which fellows discussed some current Yale issue or heard one of their colleagues describe his or her current academic work. Quite a few associate fellows attended those meetings, some of them old guard alumni still living in Yale's homogeneous past. I recall one white-haired alumnus whose florid face suggested a lifetime of Saturday afternoons tailgating at Yale Bowl. He looked as if he didn't have much use for the newfangled master.

The Fellows' Lounge was a trapezoidal room hung with Branford photographs and artifacts, which adjoined a chapel at the base of Harkness Tower where weddings and other religious services were held. On Friday nights Yale's Jewish community used the chapel for its Sabbath observance. But on one Thursday evening in October, when Branford's fellows were meeting, the Jewish congregation next door seemed to be celebrating a special holiday; through the connecting door I could hear shouts and whoops that didn't sound liturgical. I would later learn that it was Simchat Torah, a joyful festival marking the annual completion of the reading of the Torah, at which Torah scrolls are removed from the ark and carried or danced around the synagogue seven times.

I had made a new friend in Rabbi Arnold Jacob Wolf, the recently appointed chaplain of Yale's Jewish community. I liked him for his wisdom and his warmth, and he saw me as a fellow pastor to the young. What I assume happened on that Thursday night was that Rabbi Wolf, lacking a synagogue where he could dance with the Torah scrolls, led his celebrants around Branford court. Then, instead of returning to the chapel by way of its front door, he decided to go through the Fellows' Lounge where Branford's fellows were meeting.

I heard the door open and saw Rabbi Wolf standing there. Behind him was a sea of commotion and noise. He gave me a nod of acknowledgment as the host and then walked across the Fellows' Lounge and through the door leading back into the chapel, followed by the most ragtag procession I ever saw. Disheveled as gypsies in their merrymakers' garb, chanting and banging on drums, the revelers kept coming and coming; it was a mini-eternity before the last celebrant of Simchat Torah crossed the room and disappeared through the chapel door.

I took Arnie Wolf's visit as a gift of trust; he knew I would be pleased to be included in a ritual so important to the Jewish year and its cycle of worship, and I was. Other fellows were less pleased; they looked as if a tropical squall had blown through the room. On the florid face of the old Yale tailgater I caught a look that was nearer to horror; he had seen not only the future but the present. The next day, I was told, he and several other fellows called President Brewster to protest the outrage. But I never heard anything from Brewster. I think he would have enjoyed the show.

The college I inherited was cliquish and dyspeptic. One event that helped to bring us together as a community was a master's tea, held every Wednesday afternoon in the master's house, at which an invited guest talked informally to the students about his or her work. The guests were mainly men and women who had succeeded in the outside world—people whom Yale students wouldn't normally meet. Most of them were recruited from my former life in New York. Some were writers and poets and editors: Nora Ephron, S. J. Perelman, Roger Angell, Allen Ginsberg. But mostly they were public officials, sports figures, heads of businesses or ad agencies, cartoonists, filmmakers, artists, actors, musicians, Broadway producers, television producers, photographers, scientists—a mixed bag of achievers. Some were also homegrown: Yale professors and curators and coaches from obscure corners of the university.

Starting time was 4:15. On the long dining room table Mrs. Czegledi, our tenacious Hungarian housekeeper, had arrayed gallons of tea and vast quantities of triangular tuna fish sandwiches and chocolate-chip cookies. It stuns the mind to think how many thousands of those two staples Mrs. Czegledi made over the years, but every Wednesday the students fell on them as if they were rare delicacies, storing up nutrients to recharge the flagging late-afternoon brain.

At 4:30 we went into the living room, which, having been jimmied into the master's house when it was carved out of Harkness Quadrangle, was smaller than other masters' living rooms, requiring most of the students to sit on the rug. The guest sat in a chair next to a window, framed by the Gothic courtyard outside, the students literally at his or her feet. The format was question and answer, and I liked the quality of engagement on both sides. The questions were thoughtful, and the guest answered each student directly; no prepared speeches. For me the questions were as revealing as the answers. I knew the students only as individuals, not as students; I had no idea what courses they were taking or what intellectual strengths had brought them to Yale. Now, as they formulated their questions and pushed the guest's answer into new terrain, I got to know their minds and their curiosities. Many of them never missed a tea, whoever the guest and whatever the subject. I can still see two extremely bright future lawyers, Ruben Kraiem and Victoria Perry, sitting on the living room rug as freshmen and still sitting there in their senior year. A lot of teaching and learning got done on those Wednesday afternoons.

I always asked the guests to begin by explaining how they got into their present work. My ulterior motive was to jolt the students out of their assumption that successful people started their career in their present field and always knew what they wanted to do. Luckily for me, most of my guests strayed into their present job by a circuitous route, often to their surprise, after many detours. The students could hardly conceive of a career that wasn't preplanned and unswervingly pursued.

Not that the teas always went as expected. One year Yale's chaplain, Bill Coffin, took a sabbatical after his years of political activism. His interim replacement, associate chaplain Philip Zaeder, another of my pastoral pals, decided to give his year an ecumenical theme, inviting men and women from different religious traditions to be ministers in residence. They ranged from the Christian writer Frederick Buechner to a Hindu nun and a Buddhist priest named Eido Roshi. Phil Zaeder asked me if I would have Eido Roshi as a guest at a master's tea, and I gladly agreed; his would be a voice we didn't ordinarily hear.

Eido Roshi arrived, dressed in robes suggestive of a priest in a Japanese scroll. His face told me nothing. I explained that the students would ask him questions about Buddhism and that the session should last an hour. The students counted on that hour; it was locked into their weekly metabolism. Eido Roshi nodded.

Fortified by tuna fish and cookies, we all moved into the living room and I led Eido Roshi to the guest's chair. He indicated that he wanted no part of the chair. Seeing a sofa across the room, he went over and sat on it and pulled his feet up under him. Now he *really* looked like a priest in a scroll, his robes gathered around him, a look of serenity on his face. The students hunched forward expectantly. I introduced our guest and invited questions.

The students rummaged around for a topic that would get us launched, and one of them asked a question sufficiently generic to sustain an hour of theological discourse. It was something like, "What is the essence of your beliefs as a Buddhist?" Eido Roshi listened to the question and then closed his eyes, presumably to meditate on the answer. This was the awaited moment of enlightenment; we were about to be recipients of the wisdom of the East.

But Eido Roshi continued to sit with his eyes closed. The silence lengthened; enlightenment was taking its time. Finally the robes stirred and the eyes opened and Eido Roshi got ready to speak. Then, addressing the room, he said, "Do you know the song, 'The Impossible Dream'?" Murmurs of "yes" rose in reply. I was thinking, Of all the songs I know, that's the one I most loathe, except possibly "Climb Every Mountain" and "You'll Never Walk Alone."

"How does it go?" Eido Roshi asked.

The students gave it a try, flinging out the dreadful imperatives—*To dream the impossible dream, To fight the unbeatable foe*—until they finally reached the end of the unendable song. Eido Roshi nodded approvingly. Then he said, "Isn't there a second verse?" Once again the students gamely rallied to try to save the day for me; I was never more fond of them. *To right the unrightable wrong, To love pure and chaste from afar.* But the second verse sputtered out and Eido Roshi closed his eyes again. We waited in silence for further enlightenment. Then the eyes opened and Eido Roshi spoke again.

"I go now," he said. He gathered his robes around him, stepped off the sofa and walked out the door.

I never tired of the dailiness of the master's job: the small personal encounters and decisions. But the job, as I had been forewarned, had many abrasions. Today a historian looking back on Yale's recent past would find the 1970s a dreary decade, short of both money and momentum. The decade began with the university's discovery—seemingly to its surprise—that it was facing a huge deficit, and the remaining years were spent on shrinkage. Successive Yale provosts spent the decade saying no to fresh initiatives.

The infrastructure, long neglected, was now too expensive to repair. At Branford College the erosions were hard to miss; my job as master swung between the ceremonial and the janitorial. Huge water pipes burst without advance notice; all four of us knew by heart the telephone number of Physical Plant. The public rooms were shabby and dilapidated, the stonework was chipped, the basement a slum almost too squalid for recreational use. Not until the presidency of Richard Levin in the 1990s did Yale find the money and the will to restore its colleges to their original luster. At Branford the job was so sumptuously done, both inside and out, that when I came back in 2005 to meet with a writing class I almost burst into tears. Not a single pipe was dripping.

The city of New Haven was in an equal state of decay—a failed dream of '60s urban renewal, its department stores abandoned, its malls deserted, its Shubert Theater closed, its chic restaurants unborn and unimagined; there wasn't a decent meal in town. Violence born in the two nearby ghettos, The Hill and Dixwell Avenue, spilled into the residential colleges, their open gates an invitation to preppy wealth. Several undergraduate women were raped, including one in Branford, and all over Yale the gates began to close. Keys were endlessly fussed over, security meetings endlessly held. Our fortresses with their picturesque moats actually began to feel like fortresses—little islands of apprehension. Black undergraduates were sometimes challenged at the locked gates of their own college.

At the helm, Kingman Brewster also hunkered down. A highly visible campus presence in his first seven years, he largely withdrew from the public glare, traveling to alumni events whose purpose was to raise money and coax back into the nest the birds that had flown away. In 1976 he resigned to become ambassador to the Court of St. James's and, overnight, something important was gone. His replacement, the former provost Hanna Gray, ran a joyless presidency for one year and then became president of the University of Chicago, in turn to be succeeded by A. Bartlett Giamatti. None of them was able to solve Yale's pathological inability to deal with its labor unions, which resulted in periodic strikes by service workers. One of them, during my tenure, shut down the dining room for almost the entire fall term. *That* was dreary.

Meanwhile our family was moving to its own cycles of change. In 1977 Amy went away to college. Caroline, having raised St. Thomas's Day School to stature and respect in New Haven, returned to New York in 1978 to become director of the Bank Street School for Children. I stayed on at Branford for one more year, but it was a year of losses and separations, including a detached retina. Only John and I (and Felix) were left in the big house, and he would be off to college in the fall.

My life badly needed stabilizing. Unlike Yale's other ex-masters, who would always have their professor's salaries and benefits and retirement plans, I had no pension or tenure or security for the future. I was worn out from 20 years of freelancing—stitching together a patchwork of writing, editing and teaching jobs. It was time to go home and get a real job. I asked the advice of some Branford seniors, job seekers themselves, and they showed me how to prepare a résumé and type it properly and make it look nice and take it to Tyco's to be duplicated.

At the end of the academic year, in June, I cleaned the family belongings out of the master's house, including John's racetrack of slot cars in the basement, and moved back to New York. My successor was the mezzo-soprano Phyllis Curtin, whose voice classes at the School of Music I had sometimes audited. Her husband was the former *Life* photographer Eugene Cook, who shot many of the magazine's handsome layouts of singers and musicians. To take over my college President Giamatti couldn't have appointed a teacher I admired more or a couple whose values were closer to mine. I left Branford with a contented heart and didn't look back.

13

"On Writing Well"

There's almost no place as dead as a residential college whose residents have left. On commencement day Branford's courtyard was always a sea of seniors in blue academic robes receiving their diplomas, surrounded by parents, grandparents, brothers, sisters, uncles, aunts, cousins and friends—all the people whose love and support had brought them to that moment. By midafternoon the new graduates were back in blue jeans, lugging their possessions to the family car, and by the next day they were gone. Overnight, a thriving community had dissolved.

It was no place to spend the summer, and as soon as everybody's school ended we made our usual retreat to the country. We had sold our house in Westhampton Beach in 1966 to look for a more compact community, one where the same families came back every summer. The one we found was Old Black Point, in Niantic, Connecticut, a point of land jutting into Long Island Sound that had been colonized in the early 1900s by WASPs from Hartford and New York. Its houses were in the shingle style common to the Connecticut shoreline, many of them formed by patching together, not always invisibly, the fishermen's shacks that were already there.

With every decade the houses looked less and less like former fishing shacks. They were enlarged and often winterized, their yards glorified with swimming pools and privet hedges. Children bicycled and tricycled safely on the roads, and three generations of families met for athletic and social combat on tennis and croquet courts fastidiously maintained by the Old Black Point Association. Only the beach was impervious to improvement, its ramshackle bathhouses and flaking boardwalks a proclamation of WASP reserve and Hartford reluctance to spend an unnecessary nickel of insurance wealth. One old gentleman, whose vast holdings of Travelers stock were occasionally reported in the *Hartford Courant*, lit his house with 25-watt bulbs that bathed its rooms in a pale orange glow.

The house we bought was one of four built in 1916 on a hill outside the main compound by families from Hartford. Despite our arrival the incumbent families continued to call it "Hartford Hill." The four houses shared a view across a meadow that sloped down to a tidal creek full of wading birds and, beyond it, an IMAX view of Long Island Sound. Our house, a plain shingle box furnished with domestic objects of the kind now seen on *Antiques Roadshow*, suited our simple summer needs. It gave Amy and John a base for making friendships that lasted into their adult years, and it would also become a homestead for their own children, who learned to bicycle on the same deck and fell off just as often. After we took up residence in Branford College the house on the hill was our only real home and the keeper of our family memories. A Raggedy Ann doll left on a bed was likely to stay there long after its owner had gone off to college.

Once again I needed a place where I could do my writing. The only likely contender was a wooden outbuilding at the rear edge of our property, next to some woods, that the original family had built as a garage: dirt floor, four walls and a roof. It looked just big enough to hold a 1916-model car. I got a contractor to raise the shed onto cinderblocks, install a plywood floor and a few windows, and run an electrical line from the main house, enabling a light bulb to be suspended from a cord. I put my Underwood typewriter and its green metal table under the light bulb, along with a chair, a wire wastebasket, a ream of yellow copy paper and my *Webster's* dictionary. A plain wooden table completed the amenities. If I was needed on the phone, whoever was in the house, which was more than hollering distance away, would summon me with a schoolmarm's bell that was kept handy for that purpose. Later I bought a large fan; the office became fiercely hot by midafternoon, often requiring me to knock off and go for a compensating dip in Long Island Sound, which was as cold as the office was hot. Even the jellyfish stayed away until August.

"You ought to write a book about how to write," Caroline said in June of 1974 when I was complaining that I had nothing to write about. Her suggestion came out of nowhere—I had never thought of writing a textbook—but it felt right. By then I had been teaching at Yale for four years, and I liked the idea of trying to capture my course in a book. But many questions occurred to me. Who would I be writing for? What tone should I adopt? How would my book differ from all the other books on writing?

The dominant manual at that time was *The Elements of Style*, by E. B. White and William Strunk Jr., which was White's updating of the guide that had most influenced *him,* written in 1919 by his English professor at Cornell. My problem was that White was the writer who had most influenced *me*. His was the style—seemingly casual but urbane and wise—that I had long taken as my own model. How could I not agree with everything he said about language and usage in *The Elements of Style*? He was Goliath standing in my path.

But when I analyzed White's book its terrors evaporated. *The Elements of Style* was essentially a book of pointers and admonitions: Do this, don't do that. As principles they were invaluable, but they were only principles, existing without context or reality. What White's book didn't teach was how to apply those principles to the various forms that nonfiction writing can take, each with its own requirements: feature writing, interviewing, travel writing, science writing, memoir, sports, criticism, and humor. That's what I taught in my course, and it's what I would teach in my book. I wouldn't compete with *The Elements of Style*; I would complement it.

That decision gave me my pedagogical structure. It also finally freed me from E. B. White. I saw that I was long overdue to stop trying to write like E. B. White—and trying to *be*

E. B. White, the sage essayist. He and I, after all, weren't much alike. He was a passive observer of events, withdrawn from the tumult, his world bounded by his office at *The New Yorker* and his house in rural Maine, the prisoner—as his collected letters would reveal—of a thousand imagined ailments. I was a participant, a seeker of people and places and change and risk. Now I had also become a teacher, my world enlarged by every new student who came along. The personal voice of that teacher, not the literary voice of the essayist, was the one I wanted narrating my book.

For that I would need a new model—a writer I wouldn't emulate for his subject but for his turn of mind, his enjoyment of what he was teaching. Such a book wouldn't come from an English professor, squeezing the language dry with grammatical rules and regulations. It would have to come from a different field—and it did. My model for *On Writing Well* was *American Popular Song: The Great Innovators, 1900–1950,* by the composer Alec Wilder.

Wilder's book, recently published, was one I had been waiting for all my life, the bible that every collector prays someone will write in the field of his addiction. I was a collector of songs—the thousands of Broadway show tunes, Hollywood movie songs and popular standards written in the 40-year golden age, from *Show Boat* in 1926 to the rise of rock in the mid-1960s. As a part-time jazz pianist I thought I knew them well—the oldest of old friends. Alec Wilder showed me I didn't.

To write his book Wilder examined the sheet music of 17,000 songs, selecting 300 he felt the composer had pushed into new territory. Along with his text he provided the pertinent bars of music to demonstrate how the composer had written a passage he found original or somehow touching. But what I loved most about Wilder's book went beyond his erudition. It was his total commitment to his enthusiasms, as if to say, "These are just one man's opinions—take 'em or leave 'em."

Commenting on Irving Berlin's "Let's Face the Music and Dance," the sinuous beauty that wraps up Fred Astaire and Ginger Rogers's *Follow the Fleet*, Wilder writes "that it's a very well written song, that the lyric is particularly 'civilized,' that it is uncluttered, makes every point it sets out to make, indeed, deserves high praise. Yet I must admit to irritation in the presence of this form of minuscule melodrama. I call it Mata Hari or canoe music."

Wilder's pleasure, however, is to praise. That connected with my own principle of not teaching by bad example. I may cite some horrible example of jargon or pomposity to warn against the prevailing bloatage, but I don't deal in junk. If writing is learned by imitation I want my students to imitate the best.

Thus I saw from Wilder's *American Popular Song* that I might write a book about writing that would be just one man's book. I would write from my own convictions—take 'em or leave 'em—and I would illustrate my points with passages by writers I admired. I would treat the English language spaciously, as a gift waiting for anyone to unwrap, not as a narrow universe of syntax. Above all, I would try to enjoy the trip and to convey that enjoyment to my readers.

Who would those readers be? I assumed that I would be writing for a small slice of the population: college and high school students, teachers, newspaper editors, aspiring writers. I had been given tentative approval by George Middendorf, head of the college division of my longtime publisher, Harper & Row. It never occurred to me to offer it to the trade division, which had published five of my earlier books. This was presumably a textbook I was writing.

And yet I didn't think of myself as a textbook writer. I just sat down and wrote my own book, trusting my readers to tag along. I wrote in the first person ("I"), starting in the very first sentence—highly un-textbook-like behavior—and almost immediately I found myself talking to the reader directly ("you'll find," "always remember," "try not to"). It was a teacher's style, not a writer's style, and I fell into it naturally, confident that my advice would be helpful. I didn't study any other books about writing. I just went out to my Underwood every morning and worked from the notes I had used in my class.

I began by writing brief chapters on fundamental principles such as clarity, simplicity, brevity, usage and the elimination of clutter. Then I settled into the heart of the book—longer chapters explaining how to write a lead, how to write an ending, how to conduct and construct an interview, how to write about travel and science and technology and sports and the arts, and how to write topical humor. Throughout I provided examples of writing I admired in those fields. The authors I chose were very different in personality and style, but *all of them wrote good English.* That was the premise I wanted to establish: that nonfiction is hospitable to an infinite number of voices. Any style is acceptable if the writing is clear.

My only concern was that I would go broke paying for permission to reprint all those excerpts. But then I consulted the "fair use" provision of the copyright law and found that an excerpt of 300 words or less—in a book-length work—is considered fair use and may be used without payment. That rule was not only a financial lifesaver; it was the breakthrough that gave the book its pace and its rhythm. I saw that I could put the 300-word limit to my advantage. As an editor I knew that almost anything can be cut to 300 words; the material is somewhere in the marble, waiting to be quarried out. Therefore I deliberately selected passages that made a coherent point in less than 300 words and also preserved the author's style and personality. Only in a few cases did I let an excerpt run longer because the writer needed an amplitude I didn't want to violate. That decision saved the book from looking and feeling like an anthology of recommended readings. It was my book; I was the tour guide.

At that time nonfiction was still a man's world; women mainly plied the quieter waters of invented truth—novels and short stories. But the growing feminist movement empowered women writers to believe in their own reality, and they were starting to create a bold new literature of memoir, biography, and social and political concern. Only one, however, had grabbed my attention as an important long-form journalist: Joan Didion. Her book of collected magazine pieces, *Slouching Toward Bethlehem*, was just the kind of writing I was trying to teach—personal, observant, engaged—and it had worked well for my Yale students. Now, compiling my book, I used excerpts from two of her chapters. One of them began:

> This is the California where it is easy to Dial-A-Devotion but hard to buy a book. This is the country of the teased hair and the Capris and the girls for whom all life's promise comes down to a waltz-length white wedding dress and the birth of a Kimberly or a Sherry or a Debbi and a Tijuana divorce and a return to hairdressers' school.

But otherwise it was a lineup of white males—mostly the same old lions who had influenced my generation of nonfiction writers (H. L. Mencken, George Orwell, Joseph Mitchell, Alfred Kazin, E. B. White, Alan Moorehead, Norman Mailer, Red Smith). I also included three scientists who wrote with clarity and warmth (René Dubos, Loren Eiseley and Lewis Thomas); an architect (Moshe Safdie); a film critic (James Agee); a music critic (Virgil Thomson); and other favorite stylists (Garry Wills, V. S. Pritchett), all highly respectable. But a few outlaws also sneaked into the tent. One was Hunter S. Thompson, whose *Fear and Loathing in Las Vegas,* a narcotic salute to the departed Sixties, had recently been published. Another was Richard Burton, the Welsh actor. Both illustrated my point about the boundless hospitality of the nonfiction form. I also didn't think they would turn up in any other books on writing.

Here's Thompson stopping by the roadside in his drive across Nevada to cover the National District Attorneys' Conference on Narcotics and Dangerous Drugs. The irony of that particular reporter covering that particular meeting is almost a definition of irony:

> Luckily, nobody bothered me while I ran a quick inventory in the kit-bag. The stash was a hopeless mess, all churned together and half-crushed. Some of the mescaline pellets had disintegrated into a reddish-brown powder, but I counted about 35 or 40 still intact. My attorney had eaten all the reds, but there was quite a bit of speed left . . . no more grass, the coke bottle was empty, one acid blotter, a nice brown lump of opium hash and six loose amyls . . . Not enough for anything serious, but a careful rationing of mescaline would probably get me through the four-day Drug Conference.
>
> On the outskirts of Vegas I stopped at a neighborhood pharmacy and bought two quarts of Gold tequila, two fifths of Chivas Regal and a pint of ether. . . . The druggist had the eyes of a mean Baptist hysteric. I told him I needed the ether to get the tape off my legs, but by that time he'd already rung the stuff up and bagged it. . . .

There's a man writing for himself and not worrying about whether the reader is clucking his tongue; every writer should be so confident. As for Richard Burton, I came upon an article in which he described his native game of rugby. Its second sentence was one of the longest I ever saw, but it was under control all the way, and it had an accelerating mirthfulness—pure joy in the rush of language. Burton's galloping sentence was proof that, finally, there are no fixed rules. If it works, that's all we need to know:

> It's difficult for me to know where to start with rugby. I come from a fanatically rugby-conscious Welsh miner's family, know so much about it, have read so much about it, have heard with delight so many massive lies and stupendous exaggerations

> about it and have contributed my own fair share, and five of my six brothers played it, one with some distinction, and I mean I even knew a Welsh woman from Taibach who before a home match at Aberavon would drop goals from around 40 yards with either foot to entertain the crowd, and her name, I remember, was Annie Mort and she wore sturdy shoes, the kind one reads about in books as "sensible," though the recipient of a kick from one of Annie's shoes would have been not so much sensible as insensible, and I even knew a chap called Five-Cush Cannon who won the sixth replay of a cup final (the five previous encounters having ended with the scores 0–0, 0–0, 0–0, 0–0, 0–0, including extra time) by throwing the ball over the bar from a scrum 10 yards out in a deep fog and claiming a dropped goal. And getting it.

My book was also heavily male in its language. The writer was always referred to as "he" or "him." So was the reader ("coax the reader a little more; keep him inquisitive"). So was every other generic type: the humorist, the columnist, the critic. I was the product of a cultural lineage that excluded women by pronoun and never gave their absence a second thought. It would be another decade before Casey Miller and Kate Swift's *Handbook of Nonsexist Writing* came along to wake us all up.

Another person who got excluded from the book was me. I was present as an authority figure, a teacher handing down teacherlike advice. But there was no mention of the writing *I* had done that made me so certain of my opinions. How had *I* dealt with the problems I was so blithely telling readers how to solve? But I sailed through half the summer keeping my own experiences out of my book.

The reason was a fear of immodesty, born of the injunction that WASPs shouldn't "make a show" of themselves. It was OK for me to explain the decisions that other writers had made, but not the ones *I* had made. Only gradually did that affectation strike me as foolish. I would find myself recalling some writing assignment that taught me a valuable lesson, and I would think, If that was so helpful to *me*, it would be helpful to other writers. So I dipped my toe in the forbidden stream, allowing myself to describe how I approached a particular article. But I never quite stopped expecting a knock on the door by the reticence police.

I also began to include excerpts from my own writing that illustrated a particular point. To justify that heresy I told myself that I knew better than anyone else what had gone through the writer's mind—better than I knew, for instance, what had gone through Tom Wolfe's mind or Norman Mailer's mind. Typically, in the chapter on how to write a lead, I used three of my magazine leads that demonstrated three ways of getting a story started, each one different but appropriate for that piece. One of those leads, which originally ran in *Life*, began:

> I've often wondered what goes into a hot dog. Now I know and I wish I didn't.
>
> My trouble began when the Department of Agriculture published the hot dog's ingredients—everything that may legally qualify—because it was asked by the poultry industry to relax the conditions under which the ingredients might also include chicken. In other words, can a chickenfurter find happiness in the land of the frank?
>
> Judging by the 1,066 mainly hostile answers that the

> Department got when it sent out a questionnaire on this point, the very thought is unthinkable. The public mood was most felicitously caught by the woman who replied: "I don't eat feather meat of no kind."

That may not be the best of all possible leads, but I'm sure it works. It keeps tugging the reader from one paragraph to the next, inexorably leading to the department's list of what may legally be put into a hot dog, which includes "the edible part of the muscle of cattle, sheep, swine or goats, in the diaphragm, in the heart or in the esophagus . . . [but not including] the muscle found in the lips, snout or ears."

Beyond its stomach-turning richness of information, that passage conveys tremendous enjoyment; many readers have told me they still remember the line about feather meat—probably the funniest quote I ever found. I wanted my book to have a natural current of humor—a reminder to writers not to take "writing" with such solemnity.

So the three months of summer raced by, the rattle of my Underwood mingling with the chatter of birds in the woods behind my shed, until Labor Day played its annual terminator role and sent everyone back to school. By then my book was about 85 percent finished, and I was pleased with its tone; it didn't feel like a textbook. One day the thought popped into my head that it might also appeal to ordinary readers, not just to students, and I told my trade editor at Harper & Row, Buz Wyeth, that I would also be submitting the book to him. I packed the manuscript up—a bundle of double-spaced pages on yellow copy paper—and mailed it at the Niantic post office.

I called the book *Writing Well*—the simplest possible statement of what it was about.

In mid-October Buz called to say that he liked the manuscript and wanted to publish it in a hardcover trade edition. That was a bolder decision than it may appear today; there was no book quite like it and no assurance that it would find a general audience. Meanwhile, the college department agreed to publish the book simultaneously in paperback for the educational market.

The only problem, Buz said, was that a book called *Writing Well* already existed, written by the poet Donald Hall. I was very sorry to hear it. But Buz adroitly cut that legal knot by adding only two letters: *on.* To call the book *On Writing Well*, he said, might even be an improvement, making it sound less oracular and more like a mere meditation on the writer's craft. So the book got its title, along with a subtitle specifying its purpose: *An Informal Guide to Writing Nonfiction.* I would write the remaining chapters when I got back to Yale, between my college tasks, sitting by the window in Branford College.

I had always made it a point of pride—further WASP decorum—to type a clean copy of everything I submitted for publication. But in this case I ran out of time—Yale's fall term was calling me back—and I sent the manuscript to Harper without the usual retyping, every page penciled with my final cuts and corrections. Another Harper editor, after reading my manuscript, wrote an internal memo confessing that she had "snooped shamelessly" among my revisions and found that

> . . . the author himself, bless him, is still a learner, too, for some of the "clutter" that he so wisely deplores sneaked into his own draft, only to be relentlessly destroyed in the revision. I don't know whether the author would agree to this, but what better teaching device than to allow the Zinsser mind to be seen at work on his own writing, making the very improvements he is urging upon his writer readers? In every case the second thought was better than the first (or fifth than fourth, or whatever the number may be). Scholars spend their lives studying poets' revisions. It would be excellent training for beginning writers, surely, if they could spend at least a few minutes studying Zinsser's!

Far from feeling threatened by such nakedness, I welcomed the chance to be my own laboratory specimen, and when the book was published it included two facing pages of my manuscript, every sentence tightened by my cuts and improvements. Those would become the two most popular pages in the book, reprinted in every subsequent edition and in dozens of writing anthologies.

The design of the book was important to me—I didn't want it to look trendy. I wasn't bringing any new news about how to write well; I was dealing with principles of clarity and simplicity and grace that went back to Shakespeare and the King James Bible, and I wanted the design to reflect those values. For the text type I chose Caledonia, a clear and welcoming face—a good teaching type—and for the jacket I recruited Nathan Garland, a graphic artist I had worked with in New Haven, admiring his lean solutions to problems that most designers would weaken with fussy ornamentation.

"This book is going to be around a long time," Nate said after reading the manuscript, "and that's the classic feeling I want to express." The jacket he created was both simple and elegant: Bembo type on pale gray paper, the letters in black and Chinese red, instantly establishing the identity that *On Writing Well* would present to the world. It helped that the book itself was small and slender—only 8½ by 5½ inches and 151 pages long. Nobody would expect it to be anything but what it was—just one man's journey.

14

Taking It on the Road

On *Writing Well* was published in the winter of 1976. It got some pleasant reviews and sold in modest numbers, matching my own modest expectations. I had no inkling that the book would change the direction of my life, taking me and my work far beyond the classrooms of Yale.

I began to get letters and calls from colleges inviting me to come and talk about writing to their faculty and students—a visit that often began with a lecture that was open to the whole town. Almost all the colleges were institutions I had never heard of. I remember one phone call from an earnest-sounding professor at what sounded like Seminal College. I assumed that she taught at one of those English departments that were then fawning over Foucault and Derrida, and I told her I didn't do seminal. "No, not Seminal!" she said. "Semin*OLE*! It's Seminole Community College, in Sanford, Florida." I told her I could do that, and when I went there I got my first exposure to how seriously community colleges take the teaching of writing.

Deciding which invitations I could find time to accept, I favored colleges in parts of America where I had never been. Today their names come back to me in a kaleidoscope of memory, most of them small liberal arts colleges, often of religious origin, in relatively small towns, their Gothic buildings with names like Old North woven into the landscape of town life. Some of those names are Whitworth College in Spokane, Washington; Centenary College in Shreveport, Louisiana; Gustavus Adolphus College in St. Peter, Minnesota; Millikin College in Decatur, Illinois; Ohio Wesleyan College in Delaware, Ohio; Denison College in Granville, Ohio, and Kentucky Wesleyan College in Owensboro, Kentucky. Today I often think of those small campuses—safe havens of good teaching and emotional support—when I hear of high school students crazed with anxiety over getting into Harvard or Yale.

In memory's cavalcade I also revisit big universities with teeming modern campuses: Brigham Young in Utah, Boise State in Idaho, Wright State in Dayton, Towson in Baltimore, the University of Wisconsin in Madison, the University of Arizona in Tucson, the University of Southern Indiana in Evansville, Southeastern Missouri State at Cape Girardeau, crouched behind an enormous levee that protected it from—and completely blocked its view of—the Mississippi River. On all those campuses I found professors dedicated to the nuts-and-bolts labor of teaching students to write well—probably better, I guessed, than the job was done at Ivy League colleges, where writing was the neglected waif in the palace of learning. Many of the teachers who invited me to their colleges would remain friends long afterward, writing to catch me up on their lives or stopping by when they came to New York.

The huge bonus of those travels was to put me in touch with the readers of *On Writing Well.* Those readers told me which parts of the book they found most helpful and what topics they hoped I would include in future editions. What they liked most was that I made myself available. They didn't feel they were hearing from a professor; they were hearing from a writer who had wrestled with the same problems *they* were facing. They also liked the book's humor. Students, especially, couldn't believe they had been assigned a textbook that actually made an effort to keep them amused.

So I was persuaded that my initial fear of immodesty had been misguided. The best teachers of a craft, I saw, are their own best textbook. Students who take their classes really want to know how they do what they do—how they grew into their knowledge and learned from their mistakes. Thereafter, in every edition of the book I wrote more revealingly, trusting my readers to trust me if I wandered down some personal trail to illustrate a point.

It now occurs to me that I didn't really find my style until I wrote *On Writing Well,* at the late age of 52. Until then my style more probably reflected who I wanted to be perceived as—the urbane columnist and humorist and critic—than who I was.

Only when I started writing as a teacher and had no agenda except to be helpful did my style become integrated with my personality and my character.

When I prepared a second edition, in 1980, I answered those early readers' questions and added some points that had since occurred to me. I updated topical references and matters of usage, adding a section on jargon—a plague that many teachers told me they found troublesome. I expanded the chapter "Sports" to note the darker forces that had begun to corrupt that once-sanitized world, illustrating the change with a passage from Bill Bradley's book *Life on the Run*, which chronicled his seasons with the New York Knicks and the modern athlete's "struggle to stay in touch with life's subtleties." Surveying the altered landscape, Bradley says, "It wasn't my idea for basketball to become tax-shelter show biz."

At the request of many readers I wrote a new chapter, "Writing in Your Job," for all the people who need to do a certain amount of writing in their work for businesses, banks, law firms, government agencies, school systems, health care systems and other institutions. Much of that writing is pompous and impenetrable, damaging the organization it represents. My chapter tried to explain that institutions can be made human.

I also expanded the chapter "Humor", which had dealt only with the uses of topical humor to make a serious point—the kind of writing I did for *Life*. But I had since taught a humor-writing course at Yale that situated American humor in its longer historical context. Now I provided that history, aided by passages by such pioneers as George Ade, Ring Lardner, Donald Ogden Stewart, Robert Benchley, James Thurber and S. J. Perelman.

So the book went back out into the world stronger and more helpful—and still less than 200 pages long.

By 1985, when it was time for a third edition, I had long been back in New York, busy with writing projects that taught me many lessons. One was a long article for *The New Yorker* about a trip to Shanghai with the musicians Willie Ruff and Dwike Mitchell, in 1981, when they introduced live jazz to China. I dutifully explained everything I was absolutely sure a reader would need to know about Ruff and Mitchell and how I felt about their music. But when the article was edited by William Shawn, editor of *The New Yorker*, he cut a dismaying number of those sections, assuring me that the points I was so determined to make were implicit. I was nervous, but Shawn was right—readers had no trouble getting the point.

That lesson would strengthen everything I wrote thereafter. I learned to delete every sentence or phrase or word that told readers something they had already been enabled to know or were bright enough to deduce. I also tried to stop using phrases like "of course" and adverbs like "surprisingly," "predictably," "understandably" and "ironically," which place a value on a sentence before the reader has a chance to read it. Readers, I learned, are not as dumb as the writer thinks; they must be given room to play their role in the act of writing—to discover for *themselves* what's surprising or predictable or understandable or ironic. They don't want that pleasure usurped. That struck me as an important point, and I put it into a new chapter called "Trust Your Material."

I also wrote a highly personal final chapter about the values that shaped my own writing, starting with the ethical values of my parents. Called "Write as Well as You Can," it stated my credo that a writer must set higher standards for his work than anyone else does—and must defend what he writes against every editor or publisher or agent who tries to distort or dilute it.

But the big revolution since the previous edition was in the technology of how writers did their writing. Overnight, they had been given a machine called a word processor that wrenched them away from the bone-deep process of putting words on paper, and they were fearful about making the leap. Nobody was more fearful than I. But I forced myself to go out and buy one of the new contraptions, an early IBM model, and to puzzle out its arcane commands ("initialize diskette"), discovering that it miraculously eased the drudgery of writing and rewriting and retyping. That called for a new chapter.

The three new chapters ran at the end of the third edition, adding substance to the book without pulling its fabric apart.

By 1990, however, the country had changed considerably. *On Writing Well* was a child of the 1970s. I knew that its principles were still valid. But what about its references and its tone? Would it strike a new generation of readers as an old fogey's book? I took a close look and saw that my 14-year-old was slowly slipping out of touch. Without a major overhaul it would wither and die.

Most obviously, much of the nonfiction that I now admired was written by women. Yet my excerpted passages were still mostly by men—the graybeards who had been models for *my* generation of journalists, now gray-bearded themselves. The book's language was also lopsidedly male; "he" and "him" were still the prevailing pronouns, though women readers often chided me for referring to the reader as "he," pointing out that *they* did much of the nation's reading and resented having to picture themselves as men.

I began by hacking at the pronouns. I found more than 100 places where I could eliminate "he," "him," "his," "himself" and "man," either by switching to the plural or by altering some other component of the sentence. Then I took another look at all those male writers whose work I had excerpted. Some of them no longer served my purposes and were gently eased overboard. I wrote a new chapter, on memoir, the glamorous new belle of the nonfiction form, and that provided a natural habitat for passages by such newly fledged memoirists as Eudora Welty (*One Writer's Beginnings*), Patricia Hampl (*A Romantic Education*) and Vivian Gornick (*Fierce Attachments*).

Other strong women tumbled into chapters that almost seemed to be waiting for them: the nature writer Diane Ackerman, the science writer Dava Sobel, the Texas regional writer Prudence Mackintosh, the movie critic Molly Haskell, the literary critic Cynthia Ozick. All of them came bearing new sensitivities that gave the book an emotional tenor it had lacked. Many of them also brought new information. Janice Kaplan, a former student of mine at Yale, had carved a journalistic beat out of the huge gains made by women in physical stamina and athletic performance, and I gratefully added to my sports chapter two passages by her that had originally run in *Vogue*. One explained that women had always been banned from the Boston Marathon because it was generally assumed that they couldn't run 26 miles. But in 1972 Nina Kuscik broke that sexist barrier and became the first winner of the women's division:

> Those of us who knew about it felt a surge of pride, mixed with a tinge of embarrassment because her time of three hours and ten minutes was more than 50 minutes slower than the best men's times. Fifty minutes. That's an eternity in racing lingo . . .
>
> This year [1984], for the first time, the women's marathon will be an Olympic event. One of the top contenders is likely to be Joan Benoit, who holds the current world women's record—two hours and 22 minutes. In the dozen years since the first woman raced in Boston, the best women's times have improved by about 50 minutes. Another eternity. Meanwhile men's times have improved by only a few minutes. This should begin to answer the question of training vs. hormones. Are women slower and weaker because of built-in biological differences, or because of cultural bias and the fact that we haven't been given a chance to prove what we can do? . . . What matters is that women are doing what they never dreamed they could do—taking themselves and their bodies seriously.

Another woman writer, Kennedy Fraser, reviewing a book about the abused girlhood of Virginia Woolf, a tireless writer of journals, diaries and letters, revealed that what Woolf wrote in those intimate forms has been crucially helpful to her and other women contending with similar demons of loneliness and pain:

> It was the private messages [of Woolf and other diarists] that I really liked—the journals, and letters and autobiographies and biographies whenever they seemed to be telling the truth. I felt very lonely then, self-absorbed, shut off. I needed all this murmured chorus, this continuum of true-life stories, to pull me through. They were like mothers and sisters to me, these literary women, many of them already dead; more than my own family, they seemed to stretch out a hand.
>
> I had come to New York when I was young, as so many come, in order to invent myself. And, like many modern people—modern women, especially—I had catapulted out of my context . . . The successes [of the writers] gave me hope, of course, yet it was the desperate bits I liked best. I was looking for directions, gathering clues. I was especially grateful for the secret, shameful things about these women: the pain, the abortions and misalliances, the pills they took, the amount they drank. And what had made them live as lesbians, or fall in love with homosexual men, or men with wives?

That voice of vulnerability, stunningly honest, had never been heard in the alpha world of *On Writing Well.*

The humor chapter was also stuck in the dark ages—the humor of domestic mishap and male befuddlement—and I added passages by current writers like Garrison Keillor and Woody Allen who were tilting at modern quandaries and neuroses. Elsewhere I eliminated sections that time had passed by. It was no longer necessary to tell writers how to use a word processor; in only five years the population had become computer savvy.

Finally, I wrote a new chapter called "A Writer's Decisions." I had found that for many writers the crippling problem is not how to write, but how to *organize* what they have written. Yet that skill is seldom mentioned in writing classes and almost never taught. My chapter, strictly pedagogical, using one of my own articles as a lab specimen, analyzes sequentially the big and small decisions that went into a magazine piece about a trip that Caroline and I made to Timbuktu to look for a camel caravan in the Sahara. Many teachers have told me that it's more than ordinarily helpful because it puts readers into the mind of the writer during the process of creation.

When that fourth edition was published, in 1990, *On Writing Well* had sold a half-million copies, almost invisibly, with little promotion. Bigger spurts were still to come. But none of them would have been achieved without the periodic tune-ups that saved the book from the fatal sin of not keeping up with its times.

Demographically, as always, America refused to hold still. By 1994 a tidal wave of immigration had begun to reshape the national character. Around me, the neighborhoods of New York were suddenly a tapestry of exotic faces, clothes, languages, shops, signs, foods, sounds, smells and ceremonial customs. *On Writing Well* could no longer overlook those lively new Americans, and in the fifth edition I added a half-dozen passages by writers from other cultural traditions. They fell seamlessly into the chapters on memoir and writing about a place.

One passage, from *The Woman Warrior*, by Maxine Hong Kingston, the daughter of Chinese immigrants in Stockton, California, describes the acute shyness and embarrassment of being a child starting school in a strange land. Another, from *Halfway to Dick and Jane: A Puerto Rican Pilgrimage*, by Jack

Agüeros, recalls the writer's boyhood in an urban neighborhood where several ethnic principalities existed within a single block, all fiercely defended. Another, *Back to Bachimba*, by Enrique Hank Lopez, begins like this:

> I am a *pocho* from Bachima, a rather small village in the state of Chihuahua, where my father fought with the army of Pancho Villa. *Pocho* is ordinarily a derogatory in Mexico (to define it succinctly, a *pocho* is a Mexican slob who has pretensions of being a gringo sonofabitch), but I use it in a very special sense. To me that term has come to mean "uprooted Mexican," and that's what I have been all my life. Though my entire upbringing and education took place in the United States, I have never felt completely American, and when I am in Mexico I sometimes feel like a displaced gringo with a curiously Mexican name—Enrique Preciliano Lopez y Martinez de Sepulveda de Sapien.

Not all my writers from other cultural origins were born in other countries. Some grew up in the United States but felt no less like outsiders in white America. One was James Baldwin, whose *The Fire Next Time*, an electrifying account of his years as a boy preacher in Harlem, I still remembered 30 years later. Another native-born outsider, Lewis P. Johnson, a great-grandson of the last recognized chief of the Potawatomi Ottawas, describes his quest for his lost identity in an essay called "For My Indian Daughter," which begins:

> One day when I was 35 or thereabouts I heard about an Indian powwow. My father used to attend them, and so with great curiosity and a strange joy at discovering a part of my heritage,

I decided the thing to do to get ready for the big event was to have my friend make me a spear in his forge. The steel was fine and blue and iridescent. The feathers on the shaft were bright and proud.

In a dusty state fairground in southern Indiana, I found white people dressed as Indians. I learned they were "hobbyists," that is, it was their hobby and leisure pastime to masquerade as Indians on weekends. I felt ridiculous with my spear, and I left.

It was years before I could tell anyone of the embarrassment of this weekend and see any humor in it. But in a way it was that weekend that was my awakening. I realized I didn't know who I was. I didn't have an Indian name. I didn't speak the Indian language . . . I dimly remembered a naming ceremony (my own). I remembered legs dancing around me, dust. Where had that been? Who had I been? "Suwaukquat," my mother told me when I asked, "where the tree begins to grow."

That was 1968, and I was not the only Indian in the country who was feeling the need to know who he or she was.

Those immigrant and minority writers filled still another hole in *On Writing Well.* With my safely chosen samples of writing from my own culture I had undoubtedly suggested that the only writers worth emulating were white people leading mainstream lives. Now I wanted to tell Americans of every ethnic origin that their own narratives were no less valid and that they could use forms like memoir to contend with the pain of adjusting to a new homeland.

Again, I dropped outdated sections and authors and added current examples of good writing. One was a brilliant passage in Tom Wolfe's book, *The Right Stuff,* describing Muroc Field, in the Mojave Desert, the only place sufficiently desolate—"it looked like some fossil landscape that had long since been left behind by the rest of territorial evolution"—for the Air Force to use when it set out to break the sound barrier. The humor chapter got a further infusion of new blood from Nora Ephron ("Living with My VCR"), Ian Frazier ("Dating Your Mom") and John Updike ("Glad Rags"), a long-overdue shiv slipped between the ribs of J. Edgar Hoover. I also called attention to some injurious trends in the practice of journalism, such as the fabrication of quotes and the self-aggrandizement of sportswriters at the expense of the sport they are supposed to be covering.

The chapter "Science and Technology" was also showing its age. The writing was still fresh, but the science wasn't. Partial rescue came from a *Scientific American* article "The Future of the Transistor," by Robert W. Keyes. Written with simple linear clarity, it began:

> I am writing this article on a computer that contains some 10 million transistors, an astounding number of manufactured items for one person to own . . . Scholars and industry experts have declared many times in the past that some physical limit exists beyond which miniaturization could not go. An equal number of times they have been confounded by the facts. No such limit can be discerned in the quantity of transistors that can be fabricated on silicon, which has proceeded through eight orders of magnitude in the 46 years since the transistors were invented.

What student born in the transistor age, cocooned in his or her own world of microtechnology, wouldn't enjoy that passage, both as information and as a model of warm science writing?

But my big break came one day when I saw in the *New York Times* that the winner of the National Magazine Award for 1993 in the coveted category of reporting, defeating such traditional champs as *The New Yorker*, the *Atlantic*, *Newsweek* and *Vanity Fair*, was a magazine called *I.E.E.E. Spectrum.* I don't think I was the only reader of the *Times* who hadn't heard of it. It turned out to be the magazine of the Institute of Electrical and Electronics Engineers, a professional association with 320,000 members. I found its office in the Manhattan telephone directory and walked over and got a copy of the award-winning article, "How Iraq Reverse-Engineered the Bomb," by Glenn Zorpette.

The awards committee was right; it was a gem of investigative reporting. Constructed like a detective story, it traced the efforts of the International Atomic Energy Agency to monitor a secret program whereby the Iraqis, in violation of the agency's disclosure rules, came close to building an atomic bomb. *Spectrum*'s article focused on a technique known as EMIS (electromagnetic isotope separation) being conducted at a research center south of Baghdad called Al-Tuwaitha. For that process, Zorpette explained,

> . . . hundreds of magnets and tens of millions of watts are needed. During the Manhattan Project, for example, the Y-12 EMIS facility at Oak Ridge in Tennessee used more power than Canada, plus the entire U.S. stockpile of silver. Because of such problems, U.S. scientists believed that no country would ever turn to EMIS to produce the relatively large amounts of enriched material needed to produce atomic weapons . . .

> The discovery of the Iraqi program had much of the drama of a good spy novel. The first clue apparently came in the clothing of U.S. hostages held by Iraqi forces at Tuwaitha. After the hostages were released, their clothes were analyzed by intelligence experts, who found infinitesimal samples of nuclear materials with isotopic concentrations producible only in a calutron. [*The writer has already explained what a calutron is.*]
>
> "Suddenly we found a live dinosaur," said Demetrios Perricos, deputy head of the IAEA's Iraq action team.

Zorpette's article, which was both a work of science history and a political document, perfectly summarized the principles of science writing that had been taught earlier in the chapter. It also couldn't have been more current; Iraq and its weapons haven't been out of the news since.

After that fifth edition almost all the new material in *On Writing Well*—in the sixth edition, in 1998; the 25th anniversary edition, in 2001; and the 30th anniversary edition, in 2006—was self-generated, not written in response to external change but to changes in myself. I wasn't the same person who sat typing in a shed in Connecticut in 1976; the book and I had grown older together. Books that teach, if they have a long life, should reflect who the writer has become at later stages of his own long life—what work he has done and how his thinking has evolved.

I had become more interested in the intangibles—beyond craft—that produce the best writing: matters of intention and values and confidence and enjoyment. I had also done many kinds of writing that I hadn't tried before. Three were highly reportorial books: *Mitchell & Ruff*, about jazz; *Spring Training*, about baseball; and *American Places*, about 15 iconic American sites. In those books I learned to gather hundreds of facts and to let those facts speak for themselves, unvarnished. I learned to generate emotion by getting other people to tell me things they felt strongly about, not by waxing emotional myself. I learned not to wax. Those lessons got tucked into *On Writing Well*.

More important, I returned to the classroom. In 1993, I began teaching an adult class in memoir writing at the New School, in New York, called "People and Places." Its purpose is to help men and women to go in search of who they are, who they once were and what heritage they were born into. Teaching that class revealed two immobilizing traits that I otherwise wouldn't have known about. One was structural, the other psychological.

Most people starting on a memoir, I found, can already picture the jacket of the book: their name in big type, the handsome lettering of the title and the tinted photograph of a child by the seashore holding a pail. They can also picture exactly what the book will say and how it will be constructed. Their biggest problem is how to find an agent and get it published. The only thing they haven't thought about is how to actually *write* the book. Nobody has told them they will only discover its shape and content in the process of writing it—and that the finished book won't much resemble the one they have in mind.

To focus them on the *process*, rather than on the finished *product*, I invented a writing course that doesn't require any writing. I only ask my students to *talk* about their hopes and intentions for the book and about the possible ways of getting where they want to go. That forces them to confront all the prior decisions that memoir insists on: matters of voice, tone, tense, attitude, scope, narrative and the privacy of their family and friends. How do they plan to reduce the vast jumble of memories clamoring at them to be sorted out and described? My talking cure has worked for hundreds of men and women, many in near despair over the immensity of the task they have set themselves. A new chapter got written on that subject, called "The Tyranny of the Final Product."

Teasing memories out of those bright and accomplished adults, I was also struck by how unconfident they were, how apologetic, how uncertain of the worthiness of the tale they wanted to tell, although the stories they eventually dredged up from the past almost always moved the rest of us with their powerful emotions. Women in particular felt that they needed permission to believe in their remembered truth. To give them that permission I wrote two new chapters: "The Sound of Your Voice" and "Enjoyment, Fear and Confidence."

Later, in 2006, I added a chapter on family history, a cousin of memoir that has begun to attract boomers and older people, drawn by new technology that enables them to self-publish their saga for their children, their grandchildren, their friends and their local library or historical society. With that chapter I finally felt that I had said everything a nonfiction writer might need to know. It only took 30 years.

But the job turned out to be not finished at all. The 21st century brought with it a powerful new wind that blew apart the world my book had lived in. Overnight, paper was no longer king. Writers who had always written for newspapers and magazines that were printed once a day, or once a week, or once a month, awoke to find that "today" no longer existed. There was only "now"—a universe where people got their news and opinions on a computer screen at any time of day or night, much of it in the form of visual images or casual messages and blogs. *On Writing Well* was a print-based book, the product of an evaporating age, and would need radical surgery—for the 35th anniversary edition, in 2011, to deal with the new realities.

On Writing Well sold its millionth copy in 2000. It was a figure I could hardly believe or even imagine; I've never thought of myself as a "best-selling" author, and I'm still surprised when I hear that someone knows my name and my books. The numbers that mean the most to me are the hundreds of readers who have written or called to say how much they like the book and how much it helped them. Surprisingly often they use the phrase "You changed my life!" I don't take that to mean that they found Buddhist enlightenment or quit smoking. What they mainly mean is that I cleaned out the sludge in their thinking that had paralyzed them from doing writing of any kind—a phobia not unlike the fear of cleaning out the closets or the basement. (The hard part of writing isn't the writing; it's the thinking.) Now, they tell me, I'm at their side when they write, exhorting them to cut every word or phrase that isn't doing necessary work. Quite a few students and teachers have described classes in which they exhort each other to "Zinsserize it!" and they enclose sample pages of Zinsserized prose, heavily pruned. That, finally, is the life-changing message of *On Writing Well*: Simplify your language and thereby find your humanity.

I particularly like to hear from people who came upon the book by surprise, never having thought of themselves as writers, and were taken by its sense of enjoyment. The following letter, from a young woman in Orlando, Florida, speaks for all the voyagers whose affection for the book has kept me company on the journey:

> I am the night duty manager at a resort campground. I have never aspired to being a writer, but I was persuaded to take over the job of reporter, writer and editor of our weekly newsletter. Because of my moaning and groaning about what to write and how to write it, my boyfriend gave me a copy of *On Writing Well.* Now I'm having a real blast!

15

Processing Words

I was born in 1922 in Sloan Maternity Hospital, at Tenth Avenue and 59th Street, the last child of a New York family that already had three girls. My father didn't need to walk far to see the son he hoped would someday join him in the family shellac business; William Zinsser & Company was only half a block away, on a street that ran downhill to the Hudson River. The business had been founded by his grandfather, the first William Zinsser, who emigrated from Germany in 1849.

Sloan Maternity Hospital is long gone, but another hospital—an extension of St. Luke's-Roosevelt—has replaced it, and there, in 1995, on the same block, my granddaughter, Anne Fraser Ferreira, was born. I hurried over to welcome her and then took a stroll around the neighborhood. I could almost see my father walking across 59th Street to his office, a white-haired gentleman in a Brooks Brothers suit and a Panama hat in summer, and I could almost smell the alcohol dissolving the slabs of raw shellac from India in his factory, a 19th-century labyrinth of pipes and vats. My sisters and I knew from an early age the life cycle of the lac bug, which secretes a resinous cocoon onto the twigs of trees north of Calcutta. We never forgot that it was that humble insect that provided our privileged life and good education.

I can always remember what year my father was born in. He was eight months old in March of 1888 when New York City was paralyzed by its biggest snowstorm, the historic Blizzard of '88. His parents lived on Riverview Terrace, at the opposite end of 59th Street from the shellac factory, overlooking the East River; his father got to work by taking a horse car across 59th Street—the entire width of Manhattan island. On the day after the blizzard, while the city struggled to dig itself out, he walked across 59th Street through mountains of snow to bring back milk for his infant son from the cow that Grandfather Zinsser kept on his property next to the factory.

In my city the present is never far from the past. Sometimes, walking through Central Park, I pass a low brick structure that was built in 1870 to house the sheep that grazed on the nearby Sheep Meadow and thereby kept it mowed. In 1934 those grazers were evicted and their home was converted into a restaurant called the Tavern on the Green. But they were still in residence when my sisters and I were children and were sometimes sent to stay with our grandmother, who lived at 1 West 69th Street. Unskilled at entertaining the young, she would give us hunks of stale bread, saved for just such emergencies, and send us out to feed the sheep, which poked their snouts through a low stone fence to nibble at our childish fingers. Today their enclosure is a glitzy dining room, where a different kind of sheep come to be fed.

Sometimes I find myself walking through the Times Square area, where I held my first job, at the *New York Herald Tribune*, and saw new plays like *Death of a Salesman* and *A Streetcar Named Desire* for $1.10 in the balcony. Or I pass the old Time & Life Building, in Rockefeller Center, where the girl I had been waiting for finally came out of the Midwest and went to work for *Life*; Caroline Fraser and I are still married. Farther uptown I pass Lenox Hill Hospital, where our children, Amy and John, were born, or Mozart Hall, in Yorkville, where their babysitter, Mrs. Mockenhaupt, took them along to meetings conducted in German, or the school where Caroline ran a Girl Scout troop, or the club where Amy took dancing lessons, or the church where we all went on Sunday morning, dressed in our Protestant best, to hear the word of God.

It was to that city of memory that I returned from Yale in the summer of 1979. Eventually I would resume the freelance writer's life and would colonize still another part of town, the last of my writing places. But first, at the age of 56, I entered the American workplace.

The résumé that I made at Yale had been sent by several friends to Al Silverman, president of the Book-of-the-Month Club, who happened to be looking for a new senior editor to supervise the club's own publications, and one winter morning I took the train to New York for an interview. As soon as I met Al Silverman I knew he was a man I would like to work for. That night he called me in New Haven to offer me the job and said he would wait for me until the academic year ended. In July I joined a family of men and women who worked the same hours, knew each other's birthdays, kept pictures of their family on their desk, and lived with the assurance that they would be taken care of by a benign employer, even after they retired. That would be something new.

I already knew and admired the club's culture from having spent three months writing its 40-year history in 1966. It was then located in an industrial district of lower Manhattan, still very much the creature of its founder, Harry Scherman, a courtly man of 79 in the evening of a satisfying life. I found him presiding over a beehive of offices where editors sat reading manuscripts and a huge back room where women sat with card files and Addressograph machines, filling the day's orders. Since 1926 Scherman had been sending new books by major American and European authors—Willa Cather, George Santayana, Edith Wharton, Carl Sandburg, Thomas Mann, Virginia Woolf, Isak Dinesen, Ignazio Silone—to a country that was still intellectually provincial, thereby creating a nation of readers and gradually elevating the level of what they were willing to consume.

Now, in 1979, Scherman had been dead 10 years, his company had moved uptown to 485 Lexington Avenue, and the women with the Addressograph machines were gone, supplanted by a modern plant in Pennsylvania. It was also no longer a family business, having been sold to Time Inc. But otherwise little had changed. Silverman seemed to me to be Scherman's natural heir, a man who believed in the power of books to make a difference in people's lives. His staff—my new colleagues—were broadly educated men and women, enamored of books and knowledgeable about the previous work of the author they were reading and his or her place in the literary or political firmament.

New books were submitted to the club seven or eight months before publication, in manuscript form—giant mountains of paper, naked as the day they were born, unclothed in a jacket or any other cajoling garb. Yet the editors unfailingly went to the heart of each book, sensitive to its intentions—and also to its usefulness. "What our readers hunger for," I was told by Clifton Fadiman, a pillar of the club's board of judges from 1944 until his death in 1999, "is books that explain. William L. Shirer's *The Rise and Fall of the Third Reich* explained a whole age to us." A majority of the club's biggest sellers, Fadiman pointed out, have been books that readers found helpful.

As a writer working alone I had never given much thought to the collective act of *reading*—the cultural role that books play in a society's understanding of itself. Now, as I watched the incoming tide of books on every imaginable and unimaginable subject, hearing their virtues rigorously debated, I saw that books are social organisms, products of their historical moment and often instruments of change, altering long-frozen attitudes, like Betty Friedan's *The Feminine Mystique*, or reshaping public policy, like Ralph Nader's *Unsafe at Any Speed* and Rachel Carson's *Silent Spring*. That struck me as a good principle for writers to remember: Make your writing useful.

My main job at the club was editing the *Book-of-the-Month Club News*, the 32-page magazine that was sent to its million members 15 times a year. But Al Silverman also believed that the club should have an educational presence, and I was his willing henchman. One of his public-spirited projects was an annual series of lectures, held at the New York Public Library, in which prominent authors explained how they wrote their books in a particular genre, such as biography, memoir and religious writing. Afterward I edited each of the six series of lectures into a book, one of which, *Inventing the Truth: The Art and Craft of Memoir*, is still widely used and taught.

But Al's biggest gift to me was in the nature of a rescue. He stabilized my health and my finances after 20 years of piece-work, providing a safe place where I could reorganize my life. He was also unfailingly supportive of my own work, even when it wasn't *his* work, encouraging me to keep visiting schools and colleges to teach writing and to pursue writing projects that caught my interest.

One day I happened to mention that Willie Ruff was about to leave for Venice to play Gregorian chants on his French horn, alone at night in St. Mark's basilica, to test the acoustics that gave birth to the Venetian school of music in the late 1500s. Ruff had dreamed of making that pilgrimage since his days as a graduate student at Yale under the composer Paul Hindemith, who was obsessed with the sound of St. Mark's and its influence on the composers Andrea and Giovanni Gabrieli and their successors Monteverdi and Vivaldi. "You've got to go!" Al said, and almost pushed me out the door. It hadn't occurred to me that I could take time off to meet Ruff in Venice. But my subsequent article for *The New Yorker*, describing that night in the vast empty church, was the best piece I ever wrote.

In the early 1980s, unexpected as a meteorite, a bombshell shook all of us who were writers. It was nothing less than a revolution in the way writing was done. Paper, the writer's oldest friend, was declared obsolete, and so was the typewriter. All talk was of a new machine called a word processor, which projected a writer's sentences onto a screen, where they could be infinitely rearranged.

Some newspapers had already converted to the new technology, as I knew from a recent visit to an editor at the *New York Times.* Entering the venerable *Times* building, I anticipated the twinge of nostalgia I would feel as I walked across the chattering newsroom and heard the clacking of typewriters. But the floor was carpeted—*carpeted!*—and the room was as quiet as a cemetery. Reporters sat at terminals, writing their stories and sending them to editors at other terminals. No reporters were at the watercooler talking shop; all of them were alone with their own thoughts and their own bottle of water. That was my first glimpse of the office of the future.

But individual writers still wrote as they had always written, umbilically tied to their typewriter and paper and pencil and wastebasket. That was our holy act of creation, never to be relinquished to a machine. At least, I consoled myself, one of the people who wrote on a machine would never be me. I have a hate-hate relationship with mechanical objects; they can hear my heart accelerating and feel my hands quaver as I try to figure out how they work, and they withhold their secrets. Kind strangers help me at the Coke machine as I try to wedge a quarter into a chrome device that looks like a coin slot but isn't.

One day in 1981, Caroline said, "You ought to write a book about how to write with a word processor."

"*Me*?" I said. She, of all people, knew the vast gulf separating me from her idealized American male, Mr. Fixit. Her ancestors, Frasers from Scotland, cleared the American wilderness in the early 1800s, and some of her cousins still worked the land. Couldn't I even fill the car at a self-service pump? But as a writer's wife she saw that I would need to use a word processor sooner or later. So why not sooner? If I wrote a book describing how I learned to use the new machine—the log of an explorer in an alien land—the task might lose some of its terrors.

The next day I bit the bullet and went around to IBM. Actually it was two weeks later; I tend to put off tasks like going to the urologist and IBM. I chose IBM because it was still the blue-chip king of the industry; princelings like Dell, which would soon clamber onto the throne, hadn't even been born. I was extremely nervous as I walked into IBM's Park Avenue showroom. The equipment was immaculate, and so was Robert, the dark-suited sales manager, and so was Donna, his dark-suited assistant. They led me to a room full of demonstration models of the "Displaywriter," as IBM called its word processor. My instinct was to make a run for it. Among other things, I wasn't well enough dressed.

But Robert sat me down among the glistening units. Until then I had assumed that the word processor was one piece of equipment. Instead it consisted of five components: the keyboard, the terminal screen, an electronic box that the terminal rested on, a "toaster" with two slots to hold the two floppy disks that ran the system, and a giant printer sitting on a separate table. For a man who doesn't like machines it was a lot of machinery.

"It's really very simple," Robert said. "Donna's going to show you how it works." Donna slid a chair over next to mine and put her beautifully manicured fingers on what she called the keyboard module. She turned on the power, selected CREATE DOCUMENT from the "menu," and began to demonstrate the machine's functions. The screen became a dancing sea of revisions and repairs, and her words flew almost as fast as her fingers—strange words like "cursor" and "textpack" that were as familiar to her as "cat" and "dog," and maybe more familiar.

"Now you try it," Donna said.

"Maybe I can come back next week," I said. I mumbled something about an appointment.

"Just type something," she said.

I typed the usual something: "Now is the time for all good men to come to the aid of the party." Or maybe it was the one about the quick brown fox jumping over the lazy sleeping dog, which uses all 26 letters of the alphabet. I don't think I did "Pack my box with five dozen liquor jugs," that printer's darling, which uses the 26 letters more succinctly. It's just not something I say very often.

"You're very tense," Donna said. She was as soothing as a dental hygienist. I took off my jacket and loosened my tie; now at least I *looked* like a writer. To my surprise, the words sprang out of my fingers onto the screen; I had assumed that only Donna's fingers could operate the system. But *I* was operating the system. The words even looked like my words. I was *writing*! Mr. Typewriter had slipped over into the electronic age.

Before I could change my mind I told Robert I was ready to take the plunge, and we turned to the contract. That was one area still untouched by the new technology; the prices and the legal conditions were printed on paper, and so was the dotted line on which I was told to sign. So the deed was done, the Rubicon crossed. Originally I had meant to put the word processor in our apartment. But the five bulky units would take too much room, and I decided to put them in my office at the Book-of-the-Month Club. Robert said I could expect delivery in about two weeks. I was hoping it would take longer.

My office was on a well-traveled corridor, and the arrival of five big cartons from IBM was a highly visible event, all the more puzzling because I was obviously in no hurry to open them. When I finally got the units unpacked and installed, their multitudinous cables mercifully plugged into their proper outlets, my fellow workers stood in the doorway staring at the world of tomorrow. I was the first inhabitant of that world they had laid eyes on. Everyone else still used a typewriter.

Robert had assured me that IBM would provide "support help." Someone would come over and give me a lesson, he said, and the company also had a hotline in Dallas, with experts standing by, 24 hours a day. I knew how that worked; I had seen enough aviation movies in which the control tower talks the crippled airliner down when the pilot and the copilot are dead. ("Listen, carefully, Debbie. Do you see a little red button and next to it a knob that says 'altitude'? Good. Now what I want you to do . . .")

But Robert had also warned me not to become dependent. "Our instruction books are written so that in theory you should be able to teach yourself," he said. That was a theory I had no reason to believe in. Instructional prose is one of man's most glutinous creations, and IBM's three loose-leaf manuals were no exception. After a week of neurological disarray—sweating hands and thumping heart—I had a million questions, and I called Robert to ask what help I was entitled to. He said that the systems engineer assigned to my territory would be over in the morning to give me a lesson. Her name was Sherry. I instantly felt better. Sherry was coming! Sherry would make everything all right!

In the morning I was as fidgety as a teenager waiting for a date. When Sherry arrived and took off her coat I was taken back to another pedagogical moment, long buried in my boyhood, when my mother, in one of her attempts to make me more debonair, sent me to Arthur Murray's for dance lessons. Miss Vernon and I went round and round in a mirrored room, prisoners of the notion that every male can learn to tango.

But Sherry was a saving angel. She answered my questions and soothed my anxieties, bringing me the priceless knowledge that I was no longer alone. I had found a friend who understood me, someone I could always call, and afterward I asked when she would be coming again.

"There's something I have to tell you," Sherry said.

I've heard that phrase often enough—in life and in the movies—to know that it never augurs anything good. "I'm being transferred to another territory," she said.

"But we were just . . ."

"I know, I only heard about it this morning," she said. Sherry could see how upset I was—she was a sensitive systems engineer. "They'll assign someone else to my territory when all this gets straightened out."

I wanted to tell her it wouldn't be the same; it takes two to tango. But she gave me a firm handshake and wished me luck and walked out of my life.

So I was severed from mortal help, except for a few chats with Kathy at the hotline in Dallas ("I'm afraid your stuff has just gone into the electricity, Mr. Zinsser"), and gradually I learned to use the word processor by myself. Meanwhile I began writing my book, a few pages every day, trying along the way to keep myself amused to ease the tedium of step-by-step instruction. The chapter on pagination, somewhat to my surprise, began like this:

> Pagination! I have always loved the word and been sorry it doesn't mean all the things I think it ought to mean. Its sound wafts me to romantic or faraway worlds. I think of the great voyages that paginated the Indies. I watch the moonlight playing across the pagination on the Taj Mahal. I hear glorious music (Lully's pagination for trumpets) and I savor gourmet meals (mussels paginated with sage). I see beautiful women—the pagination on their bodice catches my eye—and dream of the nights we will spend in torrid pagination. The wine we sip will be exquisitely paginated—dry, but not too dry—and as the magical hours slip away we will . . .
>
> But why torture myself? The fact is, it's a dumb word that means just one thing: the process of arranging papers in their proper sequence and getting them properly numbered. It's something we all do, every day, almost without thought; we paginate whenever we scribble a shopping list on a few scraps of paper. So much for the romance of pagination.
>
> When you write on a word processor you don't have any papers to shuffle. But you still need to keep your papers in

some kind of order. (Some kind of order is what you kept them in when you used to spread them out on the floor.) Therefore the machine has to do the job for you. You don't have to keep telling it to paginate. It will paginate routinely, knowing that if it didn't you would go crazy. But frequently you will also need to get into the act. This happens, for instance, when you . . .

I knew from the beginning that I wanted to use the new technology not only to write my book but to get it composed. I would give my publisher my disks instead of my manuscript, saving the time and expense of having a typesetter retype what I had already typed. I would be Harper & Row's first electronic author. The book ended like this:

I'd like to think that Harper & Row is eagerly awaiting my two disks. But in their heart of hearts they wish I were bringing them a manuscript. They are nervous about the strange artifact that is about to arrive. They're in the business of publishing *writing*—and have been since 1817. Authors are not supposed to come bearing software.

But I've told them that we're in this together—love me, love my disks . . .

Tomorrow, when I deliver my book, dodging the ghosts of Herman Melville and Thomas Wolfe and other writers who walked through the streets of Manhattan looking like writers, nobody will mistake me for a member of the clan. I'll have no fat manuscript under my arm, only one small envelope. I'll take the elevator up to the office of my editor, Buz Wyeth, and hand him my disks. He will expect me to say something ceremonial, befitting the completion of a voyage across unknown seas.

"Don't bend them," I'll say.

The book, called *Writing with a Word Processor*, was published in 1983 and to my surprise sold 75,000 copies. Like *On Writing Well*, it succeeded, I think, because it wasn't a conventional how-to book; it was just one man's journey. Although it worked adequately as a manual, it worked mainly as therapy, offering hope and comfort to thousands of writers suddenly required to change the methods and habits of a lifetime. If *I* could do it, anybody could.

Three years later the book was obsolete, and hardly a typewriter was seen in the land. The revolution was over. At the Book-of-the-Month Club everyone now wrote happily on a word processor, electronically connected to everyone else. Only yesterday those converts had stood in my doorway disbelieving the future.

Never again would I be a bellwether of technological change. I reverted to my true nature, clinging to my Displaywriter, with its writer-friendly keystrokes and its sensible DOS operating system, long after it had also became obsolete. Like an anxious squirrel, I stockpiled replacement parts that IBM no longer manufactured and found dying companies in the bowels of lower Manhattan that repaired antique business machines—anything to postpone having to learn one of the still newer systems that had since been born, with their scurrying language of icons and windows, a language more visual than verbal, activated by a mouse that I was sure would scamper away at the touch of my hand.

16

Only in New York

The last of my writing places is in a part of town that only a New Yorker could love. It's on the fourth floor of a building at the northeast corner of Lexington Avenue and 55th Street, on a block densely packed with stores that nobody would call high-end, many with signs in the window offering discounts (ALL SHOES 80% OFF) or announcing their imminent demise (EVERYTHING MUST GO!). Across the street, in a two-story building that looks as if it's made of cardboard, a check-cashing store dispenses money that may well get spent at Didi's Duds, a few doors away, or, beyond it, at Belleza's Beauty Salon and Spa, which features eyebrow waxing and threading. Farther along, a health food store sells pills and potions extracted from the sea and from unsuspected realms of botany to reverse the damage of living unhealthily.

None of those establishments is useful to me as a writer. But wait! The cardboard building also has a 24-hour Korean deli, its inventory seemingly tuned to every survival need, and a copying shop, where the owner, Bill Bloxham, and his assistant, Siu Lai, amiably indulge my most misshapen requests. Upstairs is a fitness center where my trainer, Ed Irace, plies me with stretching and strengthening routines to loosen my writer's kinks and otherwise slow the march of time.

At the end of the block is a shop that claims to sell 6,000 magazines and newspapers, my salvation if someone tells me there's a review I "mustn't miss" in the *Times Literary Supplement* or if a former student calls to say he or she has an article in the new issue of *Mountain Bike,* or *Wallpaper,* or *Spin.* To the east I'm one block from a major post office, to the south I'm one block from my bank and a Barnes & Noble and a Dunkin' Donuts, and to the west I'm one block from a FedEx and a Staples that sells office supplies. My village has everything I need to do business.

I first came to the neighborhood by way of the office I rented from Bernard Geis at 128 East 56th Street, just around the corner. I had stayed at the Book-of-the-Month Club long enough to see it through one last milestone, its 60th anniversary, for which I wrote another history, *A Family of Readers*, and organized an exhibit at the New York Public Library that displayed the club's important books of all 60 years along with the important social and cultural artifacts of the moment.

But the company was fast turning corporate. A new generation had ascended to power within Time Inc.—grim M.B.A.'s who talked about "growing" the company with three-year plans and five-year plans but didn't talk much about books. Soon they would move the club across town into the Time-Life Building, where its identity would be swallowed by the corporate whale. It was time for me to get back to my own writing, and in 1987 I resigned and went looking for an office of my own—the search that led to Bernie Geis and my probationary trip down his fire pole. I moved my Displaywriter to Geis's fifth floor and never went down the fire pole again.

Actually the pole did serve one of my needs. Because the elevator only went as far as the fourth floor, that's where deliveries were made to Geis's receptionist. Instead of walking up to the fifth floor with our mail, she would put it in a basket attached to a string that descended through the hole surrounding the fire pole. One day I was pulling up the basket to get my mail when an old friend happened to drop by. Until that moment she thought I was a sophisticated man of letters.

I spent four pleasant years in that office, mainly writing two books: *Spring Training*, about the baseball camp of the Pittsburgh Pirates in Bradenton, Florida, and *American Places*, about 15 iconic sites in different parts of the country, like Mount Rushmore, the Alamo, Kitty Hawk and Appomattox. Those books reimmersed me in the discipline of extended nonfiction writing: traveling in pursuit of a story, talking to people who could bring the story to life, and making a narrative arrangement of what I found.

Outside, on 56th Street, all my personal needs were met by a store only six feet wide and eight feet deep. A yellow awning announced its principal wares: CIGARETTES, SODA, CANDY, COLD BEER, MAGAZINES, LOTTO. The sides of the awning listed further services: PHONE CARDS, METROCARDS, NOTARY, ATM. A bigger awning could have included MEDICINES; the store carried the full panoply of pain relievers, stress relievers, cough relievers, sinus decongestants and other nostrums that New York office workers religiously gulp to get from morning to night.

The store was established in 1977 by two men from Bombay, and I once asked one of them, Mohan, how many customers it gets every day; it's open from 6:00 A.M. to 8:00 P.M. He thought a moment and estimated the number at 4,000. By my own calculations that seemed about right: figure 14 hours times 60 minutes times 4 or 5 customers in any one minute. Nevertheless the owners always knew what I was there to buy. At that time my writer's fix consisted of America's two great narcotics—a Hershey Bar and a Classic Coke—and I also wanted a *New York* magazine every Monday. Seeing me walk into the store, whoever was behind the counter took the magazine down from the rack and handed it to me.

In 1991 the nation of Zimbabwe bought our building and everyone was evicted. Some of the renters on the fourth floor found a suite of offices on the 82nd floor of the Chrysler Building and invited me to join them; they said it had fabulous views of Long Island and New Jersey. I can live without fabulous views of Long Island and New Jersey. What I can't live without is the street life of Manhattan, reachable by a few flights of stairs. That ruled out the Chrysler Building and sent me back to the classified ads, where I found one that said "Ad agency seeks subtenant." I called and was told by Susan Cooper that she and her husband, Bernard Cooper, owned an advertising business, Cove Cooper Lewis, which had an office they wanted to rent. They were at 135 East 55th Street—one block away! I said I'd be right over.

Surveying their red brick building from across Lexington Avenue, I was amazed that I had never noticed it; it was unlike anything in that part of town. Only eight stories high but massively built in the Beaux-Arts style, it looked like a power plant mated with one of the buildings at Columbia University, perhaps the Economics Department. The first two floors were of rusticated limestone, topped by a decorative frieze; the remaining floors were of institutional brick. A wrought iron balcony ran around the fourth floor. The entrance had a portico, supported by two columns, with three flagpoles extending over the sidewalk. Inside, the lobby was made of marble and built to last.

As I would later learn from the New York architectural historian Christopher Gray, the building was erected in 1902 as Babies Hospital, an early work of the architects Edward York and Philip Sawyer, who later became the city's leading designers of banks. Babies Hospital was later absorbed into the medical complex of Columbia-Presbyterian Hospital, in upper Manhattan, and the building on 55th Street was converted to offices, its exterior largely unchanged. It didn't look like an office building to me; I thought I had the wrong address.

But Susan and Bernie Cooper greeted me warmly on the fourth floor and interviewed me in Bernie's corner office. It was one of the best New York offices I ever saw—high ceilinged and full of sunlight, with windows reaching to the floor on both the Lexington Avenue and the 55th Street side; beyond were the graceful balconies I had seen from the street. Diagonally across the avenue, demanding to be noticed, was another of New York's architectural surprises: Central Synagogue, a Moorish-style edifice of brown polychrome stone, its two octagonal towers and green-and-gold domes an unexpected whiff of the Levant. I thought, If only *this* was the office the Coopers were offering me!

It *was* the office they were offering me. Bernie Cooper, then in his late sixties, explained that the agency's business had dwindled and he had reluctantly decided to rent his own office and move to a smaller one in the rear. Once again I found myself an unwitting player in an age-old drama. I had watched my previous landlord, Bernard Geis, presiding over the remnants of his shriveled empire, its once-busy offices occupied by strangers. Now Bernie Cooper was staring at the same moment that founders of businesses think will never arrive.

But he and Susan were never anything but gracious, and the three of us became close friends. When Bernie dropped by my office it was to talk about books, not to weep over lost glories. He liked my shelf of reference books, and I would often glance up to see him burrowing in one of them for a synonym or a quotation or an unusual fact to brighten his advertising copy. His biggest regret, I think, was no longer having a proper place to meet his clients. Sometimes he would ask me whether I was planning any trips. I would mention a forthcoming visit to a school or a college, and on the appointed day my office would metamorphose into Bernie's office, complete with pictures of his family on my desk. Once, I returned from a trip before Bernie's meeting had ended. I sneaked back out and went for a walk.

That's the kind of building it was. We were a vertical family of men and women running small businesses, doggedly meeting deadlines and somehow meeting the rent. We would see each other in the elevator, briefly entering the rhythm of lives we could only guess at. Several small dogs were also brought to the building every day, their noses poking out of a woman's handbag, their commute almost over.

The building had only one elevator, which therefore served every maintenance need, both structural and human. Food was a frequent passenger—takeout concoctions ordered by phone from nearby delis and rushed by small Mexican deliverymen to the building's famished inmates. Sometimes they came into the elevator carrying huge metal platters of pizza, presumably for an office party, which they would balance on one extended arm in midair, over our heads, with balletic grace.

Trash bags and mops and pails also made the trip with us. So did technicians and air-conditioning repairmen and construction workers lugging sheetrock and power tools and coils of electrical cable. The building was a living organism, forever changing shape, amoeba-like, as successive owners tore down walls to pursue some misguided notion of profitable space. Restaurants of vaguely Russian descent came and went on the ground floor and in the basement. An elegant cigar club with new windows was wedged into the second floor and then never opened for business. Nobody asked why. Plaster innards from upper floors were carted out to dumpsters on 55th Street.

I enjoyed the building's makeshift bravado. I would walk to my office every morning, eager for whatever tasks or letters or phone calls or visits might come my way. For a while two magazine editors kept me busy writing articles. Maggie Simmons, editor of *Travel Holiday*, sent me (and Caroline) to Bali, to Siena, to the Greek islands, and on a tour of the Arabian Peninsula. She also sent me (alone) to England to interview the travel writer Norman Lewis, whose work I had long admired, and to Normandy for the 50th anniversary of the Allied landings on Omaha Beach, the pivotal event of my war.

Pamela Fiori, editor of *Town & Country*, which for a century had catechized its readers on the manners of the Eastern establishment, found me useful as her house WASP, a writer born into that milieu who knew its curious customs. One day she asked me to write a piece about my lifelong rebellion against my WASP heritage and my difficulty in coming to terms with it. I tried to beg off, but she was adamant.

"How many words?" I asked.

"A thousand words," she said.

"You're asking me to write the story of my life in *a thousand words*!" I said.

"Yes," she said. She had no pity. I went and tried to write the story of my life in 1,000 words, but it came to 1,800 words, all awkwardly born. I took it home and told Caroline I was hopelessly stuck. She went off to read it, and when she came back she said, "The good news is, it's salvageable." That's all any writer really wants to hear, and I wrestled the piece down to 1,000 words. Called "A Reluctant WASP" in *Town & Country*, it had an early passage that mentioned some of my old resentments:

> My parents had great humor and charm. In a word, they were attractive. Their house was attractive and everything in it was attractive. That was the point of being a WASP: to be at-

tractive. The laws were coded into my metabolism at an early age. Gaudy clothes and flashy cars were out. Understatement was in. A sweater the color of oatmeal was as attractive as you could get. I was careful never to be seen in a green jacket or tan shoes, or to use the wrong language. I said "curtains," not "drapes." I said "rich," not "wealthy."

Still, attractive as I was, I hated the word. "Is he attractive?" or "Is she attractive?" my mother or my sisters would ask when I talked about someone I had met. "Why don't you ask whether they're *interesting*? Or *smart*?" I would snap, crabby as an old socialist. But the word has never stopped following me around. Nor has the incessant naming of names. When I run into my WASP friends I know I'll soon hear the tinkle of tribal connections. "You'll never guess who I saw last week. Muffy Pratt! She knew your sister at Smith, and her sister Cissy was my roommate at St. Tim's. Wasn't her brother Chip in your class at Deerfield?" Even if he was, I don't admit it. I deny all memory of the people mentioned in these conversations.

But I knew I couldn't get off that easy. Anybody can complain, and many writers do, especially memoir writers, masters of retroactive blame. The hard job is to get beyond the ancient grievances and arrive at a larger point—some moment of acceptance and healing. Without such a point my piece could never succeed; it would be mere whining, not helpful to anyone else. Caroline had long urged me to stop knocking my heritage and to acknowledge its strengths, which had shaped my values. Of course she was right, and so was Pamela Fiori. It was time to grow up.

Here's how my piece ended:

> And yet . . . who am I kidding? My origins leak through every effort to conceal them. I look like an old WASP (horn-rimmed glasses) and I have the habits of an old WASP. I always wear a tie and a jacket in the city and on planes and trains. (The jacket comes from J. Press.) When I see a picture in the newspaper of a businessman without a tie, I just know I wouldn't want him handling my business. I always wear a hat. I have very few clothes. I don't own any electronic gadgetry. I drive what my wife calls "an incredibly self-effacing car." I'm punctual. I never make a scene in public. I write personal letters by hand.
>
> I'm aware that WASPs are a dying class. They are the only ethnic minority that other Americans may safely deride. But I also know that no class has so deeply imprinted its core values on the national character: honor, hard work, rectitude, public service. By today's standards of civic and corporate governance those values look good, and I'm proud to be associated with them.
>
> Today I often recognize fellow WASPs of my generation on the sidewalks of New York. They are always "nicely" dressed—old men and women facing the day with vigor and good cheer, disregarding the infirmities of age as they hurry to their next hospital board meeting or school tutoring session or fund-raising event for some underfunded worthy cause. There's something about them that's—well, attractive.

Other projects took me in other new directions. Several years went into *Easy to Remember: The Great American Songwriters and Their Songs*, the book I most enjoyed writing, about my lifelong romance with the Broadway show tunes, movie songs and popular standards known as "the great American songbook." I was born with a musical ear, and I've played those songs all my life, intimately familiar with their harmony and construction and with the lyrics of Lorenz Hart, Ira Gershwin, Johnny Mercer, Dorothy Fields, Frank Loesser and other giants of the form—men and women as much in love with the American language as I was. I've seen all the great musicals, starting when my parents deemed me old enough to be exposed to the risqué ballads of Cole Porter, and I've written magazine articles about Harold Arlen and Richard Rodgers and other songwriters, sometimes catching them at the out-of-town tryout of their latest show. (Arlen received me in yellow pajamas in his suite at the Warwick Hotel in Philadelphia the morning after I saw *House of Flowers*. Rodgers was more decorously dressed at the theater in Toronto where *No Strings* tried out.)

Everything I knew and had thought about went into *Easy to Remember*, along with photographs, sheet music covers, memorabilia, lists and archival notes. I wanted the book to have photographs of the songwriters that weren't the same old poses, and at first I thought of hiring a picture researcher. But then I thought, I'll bet most of those composers and lyricists have a widow or a son or a daughter who lives within walking distance of my office, or only a few subway stops away.

So I set out. I found Hoagy Carmichael's son, Hoagy Bix Carmichael, at Broadway and 49th Street. I found Frank Loesser's widow, the singer Jo Sullivan, on 45th Street near Sixth Avenue, and Jule Styne's widow, Margaret Styne, at Park Avenue and 73rd Street, and E. Y. "Yip" Harburg's son Ernie on Lafayette Street. Betty Comden was in an apartment high over Lincoln Center, John Kander a few blocks to the east. The daughters of Irving Berlin and Richard Rodgers—Mary Ellin

Barrett and Mary Rodgers Guettel—were on East 67th Street and Central Park West. The singer Margaret Whiting, adoptive daughter of Johnny Mercer, who had no children, was on 58th Street west of Fifth Avenue.

Most of them would briefly disappear into a closet and emerge with a tattered brown envelope stuffed with old photographs, telling me to borrow what I needed. Margaret Styne let me borrow a framed photograph on her wall that showed Jule Styne, then a piano prodigy named Julius Stein, making his debut with the Chicago Symphony at the age of eight, suitably garbed for the higher slopes of music in a suit with a ruffled collar and cuffs. One of my favorite pictures shows a tanned and handsome George Gershwin sitting at a card table in the California sun, in the summer of 1935, orchestrating *Porgy and Bess.* It was loaned to me, along with some unusual pictures of Harold Arlen, by Edward Jablonski, biographer of both Arlen and the Gershwin brothers and an almost-member of the tribe, who looked after Arlen with sonlike devotion in his declining years. Ed lived—of course—on the Upper West Side. Only in New York would the relatives of the songwriters be so easy to find, their village still intact, not far from the lights of Broadway where their songs were first belted out by Ethel Merman, Carol Channing and other forces of nature.

Music and musicians were now a major presence in my life, especially when I started playing club gigs in the mid-1990s with the cartoonist Arnold Roth, who had moonlighted as a jazz saxophonist since his own boyhood. Now, when I leave my office at the end of the writing day I might appear to be heading home. But I could just as easily be off to meet Arnie at the

Cornelia Street Cafe, in Greenwich Village, where he and I have played for so many years that the owners, knowing that their piano bench is too low for me, will have put a telephone directory on it. Brooklyn and Queens are just the right height; Manhattan is too thick.

In the 1990s I also went back to teaching, mainly at the New School, in the evening. My course is taken by women and men—mostly women—eager to recover their past and thereby try to understand its still-painful conflicts. I've found that the teacher of memoir writing has become one of society's listeners, joining the therapist and the rabbi and the priest. Outwardly, my students appear to have their lives together—they are neatly dressed and they turn up promptly every week, some from quite far away. But when they talk about the personal stories they want to write I discover that many of them are carrying a heavy load of adversity, often the residue of childhood abuse or abandonment or of their own subsequent addiction or badly broken marriages. I once thought only certain families were dysfunctional. Now I think that maybe there's no functional family. Which I guess any psychiatrist knows.

Not being psychiatrists, writing teachers have no clinical tools to treat their flock. The only tool I can offer is the mechanism of writing. I can help my students to think about how to use writing to make sense of their family narrative. I can also give them permission to believe in that story and in their right to tell it, especially women, who devalue their own experience and worth. Finally, I can provide a safe environment—a place where my students know they can trust their life story to 20 strangers.

Much of what I learned from those men and women went into my next book, *Writing About Your Life*. A teaching book posing as a memoir, it consists of 13 chapters about episodes in my past that I recall with gratitude or amusement. But I also pause to explain the technical decisions I made—matters of selection, reduction, focus and tone—that every memoir writer struggles with. Recently the waters of memoir writing have been famously roiled by authors who deliberately obscured or improved their story or fabricated it entirely—running away from the truth. But I see my students desperately trying to *clarify* their story—running *toward* the truth. I admire their honesty and their courage.

Meanwhile, back at 135 East 55th Street, the inevitable day arrived when the Coopers' lease expired. A marketing firm that had previously occupied parts of two upper floors, Susan M. Rafaj Associates, took over the fourth floor and we were dispossessed. Bernie and Susan found a suite of smaller offices on the sixth floor and took me along, still their subtenant. The view wasn't as good—I had to crane my neck to see the synagogue—but the building was home, and we pieced together five more good years. Then a team of wreckers came to knock down all the walls for some new incarnation of the space.

Waifs once again, we were given asylum on a lower floor by a consulting firm whose staff of three people sat at their computers all day in funereal silence. They offered me a small back room with no window, which I gratefully took, not wanting to leave the building. But it was a desolate year, and I began to look elsewhere. One day I climbed all eight floors of our building, ringing every doorbell to ask about a spare office. Most of the doors had a nameplate that left me no wiser about what the company actually did, nor was I further edified by a peek inside; I saw one rock garden that spoke of higher aspirations than I associated with the building. The people who opened the door looked at me with compassion and puzzlement—a hobo from some parallel universe. But none of them had an extra inch.

Just when I was at my most morose I was rescued by a merciful angel. On the fourth floor, Susan Rafaj, who knew how unhappily I was immured down below, freed a pleasant office for me in the midst of her marketing agency, just down the hall from where I first rented from the Coopers. Once again I had windows reaching to the floor, and I placed my desk where I could look at Central Synagogue, my anchoring landmark. In late afternoon the sun glinted off the gold of its twin green-and-gold domes, briefly redeeming the drabness of Lexington Avenue down below. According to a plaque on its facade, it's the oldest synagogue building in continuous use in New York City, designed by the American architect Henry Fernbach and built in 1872. But I had always wondered about its Oriental motifs and obviously symbolic shapes, which, I suspected, didn't spring fully formed from the brow of Mr. Fernbach.

One day a young writing student named Peter Katona came to my office and happened to mention that he was born in Hungary and brought to America at the age of five. Noticing the view from my window, he told me that his grandfather was the rabbi of the Dohány Street synagogue in Budapest, one of the holiest Jewish sites in Europe, which was the model for "my" synagogue on Lexington Avenue. The next day he sent me some photographs of that synagogue, which he had taken on a recent trip home to visit his grandparents. The Moorish-style exterior, including the two octagonal towers and their green-and-gold domes, closely matched what I had long been looking at.

I have six pictures on the walls of my office. One is a painting by our son, John Zinsser, from one of his gallery shows. One is a poster for my book *Easy to Remember*, a collage of old sheet music covers. Two are period photographs that I bought at antiques shows. One of them is a street scene under the elevated tracks, a latticework of sunlight and shadow, of old trolleys and automobiles and couples ambling past stores selling hosiery and eyeglasses and 88-cent shirts. The other is a baseball scene of such classical purity that it could have been staged. Against a backdrop of billboards for banks and department stores of the 1920s, Babe Ruth's still-slender body is coiled in readiness, his bat raised high behind his head. The pitcher has just completed his delivery, and the ball is in midflight, its path tracked by the catcher and by the umpire, hands clasped behind his back in the immemorial stance of judgment. Farther away, the shortstop is crouched forward expectantly, although the Babe was not known for hitting grounders to the left side of the infield.

On the wall over my computer is a poster of a limestone head of antiquity, brought home many years ago from a museum in Jerusalem. That beautiful half-smiling face has watched over me in many writing places, a link to my boyhood love of the classics and the Latin language, whose roots still fortify my own writing.

On another wall is a framed photograph of E. B. White, a gift from Jill Krementz, who took the picture at White's home in North Brooklin, Maine. The 77-year-old writer sits on a crude bench at a crude table in a small boathouse, typing on an old manual typewriter. The only other objects are an ashtray and a nail keg, obviously his wastebasket. A small window is open to a view across the water. White has everything a writer needs: paper, a writing implement, and a receptacle for all the sentences that didn't come out right. Since then writers have been given a computer to replace the typewriter and a delete key to replace the nail keg. But nothing has replaced the writer; he is still stuck with the job of saying something that other people will read. Younger writers who come to my office find their eyes gravitating to the old man pecking at his typewriter, still practicing the simplicity that he and his style always preached.

Like White's boathouse, my life is as simple as I can make it. When my IBM Displaywriter and its giant printer finally expired, having survived all my migrations up and down the building, I bought a Dell computer, which amply serves my writing needs. (The long-feared mouse was quickly tamed.) I've never activated the system for sending and receiving e-mail; I don't need e-mail in my work and I don't want to be its captive.

"But how can I *reach* you?" friends say. Their tone suggests that I have entered a Trappist monastery.

"Give me a call," I tell them. I enjoy the human pulse of a phone conversation; e-mail is a one-way street.

"But I want to send you something," they say.

"Put it in the mail," I say. The *mail*! They look as if all memory of the envelope, the stamp and the mailbox has leached out of their brain.

Every day my computer interrupts me to announce that it has upgrades ready to install. I always click on the box that says CONTINUE WHAT I WAS DOING. Isn't that what all of us want to be allowed to do, assaulted at every turn by peddlers of the latest option? As the husband of a competent woman who has spent hours of misery undoing the havoc caused by Microsoft's installation of features she didn't ask for and would never use, I consider "upgrade" a dirty word. My prayer for us all is: Please, God, let us continue what we were doing.

Outside, in my village, half the people are now connected by silicon chip to someone who is somewhere else. Their thoughts are not on Lexington Avenue, or on the pleasure of being alive in that time and place, or on the courtesies of negotiating a crowded sidewalk. They have broken the urban contract.

In this last of my writing places, all the strands of my life come together. I never know what outpost of my past I'll hear from. Most people who telephone feel that they know me from *On Writing Well*. A woman named Fatéma Al-Rasheed called from Kuwait to say that she had some writing problems she wanted to ask me about. A few weeks later she was in my office, bearing a gift of dates packed in a sumptuous wooden box. An editor from Los Angeles left a message asking me to call back as soon as possible. She and her collaborator were mired in an ethical dilemma they hadn't been able to solve—a matter of rights and credits. When I reached her she was in her car. "Are you on the freeway?" I asked. She said she was about to get on the freeway.

"Well, pull over and park somewhere," I said, and for 20 minutes we traversed the hills and valleys of literary fairness and arrived at a place where she felt comfortable.

Many younger writers have taken me as a mentor; they just look me up in the phone book. "I know how busy you are," they say, assuming that I spend every minute writing at my computer. I tell them I have many ways of being busy, and this is one of the ways I like best. I particularly like to be busy with people who want their writing to make a difference, and by now I have a small shelf of their books.

One that recently arrived was *Santiago's Children*, by Steve Reifenberg, an account of the orphanage in Chile where, as a young man, he found his life work as a social activist in Latin America. I first met him at a workshop in the early 1990s; now, in 2008, here was his book, handsomely published by the University of Texas Press, accompanied by a note that said, "You believed this manuscript would become a book long before I did."

I put *Santiago's Children* on my shelf next to *Awakening Hippocrates: A Primer on Health, Poverty and Global Service*, by Edward O'Neil Jr., an emergency room physician in a Boston hospital, who founded a nonprofit organization, Omni Med, to foster health volunteerism and ethical leadership in Belize, Kenya, Guyana and other poor nations, based on his own work as a doctor in the field.

Calling to introduce himself, Ed O'Neil explained that he was writing a book to provide information—facts not gathered anywhere else—for men and women who hoped to do health care work in third world countries. I immediately recognized the importance of the project and the commitment of the man on the phone. But the early drafts he sent me were hopelessly verbose, consisting of dense statistical chapters analyzing the causes of global poverty and violence and the huge disparities in world health care, along with historical chapters describing the work of the field's revered pioneers, including Albert Schweitzer, Thomas Dooley and Paul Farmer. Only radical surgery could save Ed's vision for the book.

Back and forth between Boston and New York, year after year, went the phone calls and drafts—a trusting friendship between two friends who never met—and I often worried that Ed would lose heart. But he turned out to be a type of writer I had never encountered or even knew to exist—someone not even faintly discouraged by the thought of throwing away huge chunks of time, research and labor. His feelings, he said, were not an issue. He wanted to learn to write well and to write a useful book, and now, in 2007, here it was in the morning mail, published by the American Medical Association. "You gave me the courage to find my own voice," he wrote on the title page.

In the fall of 2004 I got a call from Elizabeth Fishman, a dean at the Columbia University Graduate School of Journalism. She wondered if I would have any interest in tutoring foreign students who need extra help with their writing. She knew I had done a lot of traveling in Africa and Asia and other parts of the world where the students were from, and she thought that would give me a feeling for their cultural heritage. About 75 students from more than 30 countries are admitted to the school every year.

I told Dean Fishman I would very much like to try it. It was something I had never done, and it would put me in touch with interesting young men and women whom I would otherwise never meet. I wouldn't be part of the school's formal structure of grading the students; my only agenda would be to help them. If their professor thought they were in trouble they could be sent to see the doctor.

So I got still another set of constituents, and I began taking the subway to Columbia one or two afternoons a week. There I might meet for an hour with a young man from Ethiopia, then with a young woman from Bhutan and then with a man from Sierra Leone. Or it could be someone from Beijing or Thailand or Iraq or Uzbekistan. Every student was a different linguistic problem.

After several years the meetings at Columbia became too hard to arrange; the students were all over the city covering their beat and were often delayed or called to another assignment. Instead they began coming to my office—the latest in my string of unusual visitors. Nobody answering the door is surprised to see a woman wearing a burka or a Bhutanese tribal silk.

When I meet students for the first time I ask them about their country and about their family back home. I want to relax them and find the person beneath the anxiety of competing with American-born students in a school that's run like a Marine boot camp. They are enormously likable men and women. Women students from China sometimes bring me tea; a young Korean woman sent me an invitation to her wedding in Seoul the summer after she graduated—an event whose approach we had discussed. She couldn't believe how meddlesome her placid mother had become. I assured her that Korea has no monopoly on bossy mothers of the bride.

Sometimes I ask foreign students about the structure of their language, trying to glimpse how to reconfigure their thinking to *my* language: plain English. "What kind of language is Arabic?" I asked a woman from Cairo. "It's all adjectives," she said. Of course it's *not* all adjectives, but I knew what she meant: It's decorative. The last thing she needed as a journalist was a language that was all adjectives; she would have to be converted to the active verb and the simple declarative sentence. Students from Latin America, justly proud of the Spanish language's long and melodious nouns expressing elegant abstractions, had to be given the bad news that long and melodious nouns would be their ruin; those nouns have no action—no people propelling the story forward. The nouns would need to be reborn as active verbs, the sentences cruelly shortened.

Working so intently with those young men and women, both of us hunched over a yellow pad where I was trying to bring sequential order to what they were trying to say, both of us struck by how hard we had to *think*, I've been brought back to the fundamentals of language, and I find that satisfying. I've lived by those fundamentals since I began my own journey at the *Herald Tribune* a lifetime ago.

Ours was a journalistic world almost unrecognizably different from the one inhabited by the Columbia students I'm now tutoring. Every May at their commencement I hear the dean, Nicholas Lemann, warn them that they are going forth into a career that barely resembles the traditional newspaper model, where reporters wrote for one print medium that had one deadline every 24 hours. Today, he said, they would need to possess multiple skills, equally adept at writing for round-the-clock Web sites and blogs and at making and editing videos and photographs and audio recordings to accompany their stories, all of which could be instantly transmitted to an editor from any part of the globe.

And yet, stuck with my traditional skills, I'm not feeling obsolete. Language is still king, writing still the supreme conveyor of thoughts and ideas and memories and emotions. Somebody will still have to *write* all those Web sites and blogs and video scripts and audio scripts; nobody wants to consult a Web site that's not clear and coherent. Whatever new technology may come along, writers will continue to write, going wherever their curiosities and affections beckon. That can make an interesting life.

Acknowledgments

Once again I thank John S. Rosenberg, my onetime student and longtime counselor, for his wise guidance and editing skill at every stage of my journey.

For their helpful and supportive reading of the successive chapters of my manuscript I warmly thank Jane Flanagan, Douglas Goetsch, Jim Nelson, Al Silverman, John Snyder, Polly Steinway, Warren Wechsler and Caroline Zinsser.

I warmly thank Mark Singer for gathering the recollections of almost 50 men and women who, like him, were writing students of mine at Yale in the 1970s. He used their memories in a talk about my methods as a teacher, and I've drawn on some of them here in my own account of those years (Chapter 10).

I'm grateful to my editor at HarperCollins, Bill Strachan, who believed in the book and nurtured it to publication with thoughtfulness and care.

For a historical perspective on the Brewster presidency at Yale, I'm indebted to the excellent book by Geoffrey Kabaservice, *The Guardians: Kingman Brewster: His Circle and the Rise of the Liberal Establishment*, Henry Holt & Company, 2004.

Information about the origins and style of the building at 135 East 55th Street was generously provided by the New York architectural historian Christopher Gray.

Natalie Cruz cheerfully came to my rescue whenever the computer brought my book to a halt with behavior that left me mystified.